The Portfolio Diet for Cardiovascular Disease Risk Reduction

The Portfolio Diet for Cardiovascular Disease Risk Reduction

An Evidence Based Approach to Lower Cholesterol through Plant Food Consumption

Wendy M. Jenkins

Amy E. Jenkins

Alexandra L. Jenkins

Caroline Brydson

ACADEMIC PRESS

An imprint of Elsevier

ELSEVIER

Academic Press is an imprint of Elsevier
125 London Wall, London EC2Y 5AS, United Kingdom
525 B Street, Suite 1650, San Diego, CA 92101, United States
50 Hampshire Street, 5th Floor, Cambridge, MA 02139, United States
The Boulevard, Langford Lane, Kidlington, Oxford OX5 1GB, United Kingdom

Library of Congress Cataloging-in-Publication Data
A catalog record for this book is available from the Library of Congress

British Library Cataloguing-in-Publication Data
A catalogue record for this book is available from the British Library

ISBN: 978-0-12-810510-8

For information on all Academic Press publications
visit our website at https://www.elsevier.com/books-and-journals

Publisher: Stacy Masucci
Acquisition Editor: Katie Chan
Editorial Project Manager: Sam Young
Production Project Manager: Punithavathy Govindaradjane
Cover Designer: Christian J. Bilbow

Typeset by SPi Global, India

Working together
to grow libraries in
developing countries

www.elsevier.com • www.bookaid.org

Dedicated to Dr. David J.A. Jenkins.

The best way to learn is by example.
Thank you for being that example to us all.

Contents

Foreword

My daughters have asked me to write a Foreword for their diet book and it is with tremendous pride that I declare many conflicts of interest.

They saw the need to translate a therapeutic diet that lowers serum cholesterol into something that was within the grasp of ordinary people while not talking down to those with a genuine scientific curiosity into how the dietary changes work. A unique aspect of their book is that their choice of diet is not only justified on scientific grounds, but also by environmental and humanitarian considerations. The environmental challenge is increasingly a component of national guidelines, and is echoed by the Scientists Second warning to Humanity in 2015 as well as by the UN Intergovernmental Panel on Climate Change (IPCC), which has given us only 12 years to dramatically change our habits. Scientific articles (Lancet EAT) and authors (David Katz, Suzie Cameron) have taken environmental science from a discipline that nutritionists kept separate from nutrition, to one that is intimately linked.

My Daughters' book moves naturally into that space.

Wendy conceived the book and wrote the first draft. Amy contributed chapters, co-wrote sections and added comment to the rest. My wife Alexandra edited all the chapters and my sister Caroline, who wrote the first Portfolio recipe book, added her recipes to the significant number of new recipes and gave valuable criticism. It has been a family affair, and I was privileged to taste (eat) the recipes for many dinners over many weeks.

Their story starts in Chapter 1 with the demonstration of how diet can help to reduce risk factors for cardiovascular disease. They discuss a range of effective dietary strategies and show where the Dietary Portfolio fits. In Chapter 2, they discuss the Dietary Portfolio in detail and its effect compared to statin drug therapy that reduces low density lipoprotein cholesterol and other risk factors, including the inflammatory biomarker c-reactive protein that may be related to increased cardiovasular disease and cancer mortality. They, then in Chapter 3, break the Dietary Portfolio into its component parts: soy and other legumes, viscous fiber, plant sterols, and nuts, and discuss possible mechanisms of action. Chapter 4 contains valuable tips for incorporating the Portfolio components into your diet, without which there can be no Dietary Portfolio. Quite naturally, Chapter 5 contains the recipes that lie at the heart of good diet books. Finally, Chapter 6 gives what is perhaps the strongest justification for consuming the Dietary Portfolio and diets like it, namely, the environment concerns, and humane reasons, for a plant-based diet. If science does not convince you to change your diet for health reasons, then maybe your conscious will.

David JA Jenkins, OC, MD, FRSC, FRCP, FRCPC, PhD, DSc
Canada Research Chair in Nutrition and Metabolism
Professor, Departments of Nutritional Sciences and Medicine, Faculty of Medicine, University of Toronto, Toronto, ON, Canada
Director, Clinical Nutrition and Risk Factor Modification Centre, St. Michael's Hospital, Toronto, ON, Canada
Scientist, Li Ka Shing Knowledge Institute, St. Michael's Hospital, Toronto, ON, Canada
Staff Physician, Division of Endocrinology, Department of Medicine, St. Michael's Hospital, Toronto, ON, Canada

Acknowledgments

We would like to thank Dorothea Faulkner, PhD, RD, for her contribution and development of the original cookbook. We would also like to thank all the participants in the original Portfolio studies, whose dedication resulted in groundbreaking research. We would like to thank Punithavathy Govindaradjane for going back and forth with us on the various versions of the book. John Sievenpiper and Arash Mirrahimi for reading over chapters of the book and giving valuable feedback. Chef Sara Harrel for her expert advice and guidance in developing and testing recipes.

Finally, we would like to thank all our recipe testers, tasters and artistic advisors, many of whom, consumed more psyllium than we ever thought humanly possible:

Aljosha Filippov
Claire Zhang
Chris Cheng
Connie Liao
David Brydson
Djoerd Ameschot
Eli Brouwer
Jaclyn Prystupa
Jennifer Callinder
Jingjue Wang
Howard Koster
Maureen Prikken
Margot de Man
Maya Nemeth
Nicole Scott
Shane Pyman
Tansha Anand
Teresa Bicknell
Violeta Onland
William Francey
Yini Yiang

Introduction

Let food be thy medicine and medicine be thy food.

—Hippocrates 470–340 BC, Greece

LDL cholesterol has been shown to be a strong indicator for the development of coronary heart disease.

We would like to welcome you to the Dietary Portfolio. In our book, we have tried to give you the reasons why following the Portfolio Diet will improve the health of you, your family and the planet. The Dietary Portfolio is an option for those who cannot or do not wish to be on cholesterol-lowering medications or wish to improve the effectiveness or reduce the dose of CVD-related medications. As the Portfolio Diet reduces Low-density lipoprotein (LDL-C) and other CVD risk factors, it may also be helpful for those who want to be proactive and avoid the development of CVD over their lifetime. Ultimately, this diet is for anyone who wants to follow a healthier, more sustainable and compassionate diet. It is unfortunately not suitable for those with sitosterolemia (a rare condition affecting plant sterol absorption). As always, it is a good idea to check with your healthcare provider before starting a new diet.

The Portfolio diet stands out from other popular diets as it is an evidence-based diet with its ability to reduce CVD risk factors scientifically tested [1–3]. Since its development, it has been promoted internationally and is recommended by the Canadian Cardiovascular Society (CCS) [4], the European Atherosclerosis Society (EAS) [5], and Heart UK [6]. The Portfolio diet is most well known for its ability to reduce LDL-C levels [7,8]. This type of cholesterol (LDL-C) is known to contribute to the formation of fatty arterial plaque, which can eventually lead to blocked arteries, and increase the risk of heart attack and stroke.

In clinical trials, the Portfolio diet, in the context of a plant-based diet, were able to reduce LDL-C by 28%, an amount equivalent to the reductions seen by those taking Lovastatin, a statin drug used to lower cholesterol [3]. When health outcomes were evaluated using the Framingham risk equation, which uses patient history and health indicators like blood pressure, weight, and cholesterol, to calculate the risk of developing coronary heart disease (CHD), the risk for those consuming the Portfolio diet was reduced by 25%. This research was the first to demonstrate that a dietary intervention could reduce LDL-C levels to a similar extent as a first-generation statin. Longer-term Portfolio diet studies have been conducted in several locations across Canada. Unlike the first study, these studies asked the participants to find and purchase the Portfolio foods themselves. These studies have also had very positive results, with reductions in LDL-C ranging from 13% to 14% [2]. These reductions were seen even though the different locations had very different food environments where access to portfolio foods varied. Unsurprisingly, this study found that individuals in areas with easy access to foods that support the Portfolio diet like fresh fruit, vegetables, and a wide selection of meat analogs were better able to adhere to the Portfolio diet and saw larger reductions in CHD risk. The impact that different food environments can make on the success of a diet is one of the main reasons for this book. Although it's impossible to change the food environment you live in, we hope this book can help you better navigate your current environment by providing not only the scientific evidence behind the diet, but also tips, tricks, guidance, and easy-to-prepare recipes.

The protective effects of the Portfolio diet are not limited to its effects on LDL-C. Other beneficial effects of the Portfolio diet include the reduction of inflammatory biomarkers like c-reactive protein (CRP), increased levels of which have been associated with heart disease and cancer [9,10]; improved glycemic control or lower blood sugar levels after a meal; an increase in the amount of high-density lipoprotein cholesterol (HDL-C) or "good" cholesterol, in relation to LDL-C, the "bad" cholesterol, and finally blood pressure reductions similar to those seen in DASH diet, the gold standard for dietary blood pressure reduction [2,11]. Together these factors have found to contribute to a reduced risk of CVD and have shown potential in the management of type 2 diabetes [12].

The Portfolio diet is composed of four main pillars. Each of the four pillars work in tandem to reduce risk factors. The four pillars of the Portfolio diet are

1. *Plant sterols*, found in enriched margarine, juices and supplements.
2. *Viscous fiber*, which is found in foods such as oats, barley, okra, and psyllium.
3. *Nuts and seeds* like almonds, walnuts, and sunflower seeds.
4. *Plant-based protein* found in foods such as tofu, soy milks, meat analogs and legumes.

The four pillars of the portfolio diet:
plant sterols
viscous fiber
plant-based protein
nuts and seeds

The Portfolio diet was developed with the intention of acting like a stock portfolio where assets, or in this case, foods, can be collected and added as new scientific evidence emerges. In this way, it is a dynamic and ever-evolving diet. It is situated in the context of a plant-based diet for health, environmental, and ethical reasons. Users of this diet are encouraged to consume foods in the context of a plant-based diet, which contains no animal products. It is recommended to try to minimize the consumption of animal products for best results. In addition to health, sustainability of diet will also be improved with increased adherence to a plant-based diet. As the Portfolio diet consists of a variety of foods, which can be used in a variety of dishes, it can easily adapt to fit personal tastes, as well as a variety of cultural cuisines. This portfolio of foods provides a healthy base ready to be customized. For a short summary on how to incorporate the Portfolio diet into your lifestyle see infographic on the following pages, which was developed in collaboration with the original Portfolio researchers at St. Michael's Hospital in Toronto.

THE
PORTFOLIO DIET
An evidence-based eating plan for lower cholesterol

EGGPLANT

CHICKPEAS

VEGGIE BURGER

ALMONDS

SOY MILK

WHAT IS THE PORTFOLIO DIET?

The portfolio diet is a way of eating that evidence has shown can help lower cholesterol and your risk of heart disease. Instead of focusing on what you can't eat, the Portfolio diet is about what you can add to your menu!

The diet includes a "portfolio" of plant foods that you can choose from.

WALNUTS

TOFU

PLANT STEROL MARGARINE

OATMEAL

BEANS

Research shows that medications and diet both work to lower your cholesterol. Medications can be more effective and easier, but some people don't want to take medications, cannot tolerate the side effects, or want to combine a nutritious diet with medications.

HOW DOES IT WORK?

The Portfolio diet is exactly as it sounds. It takes a few dietary patterns that have been shown to lower cholesterol and puts them together. To lower your cholesterol, you can "invest" in any one pattern, or some of them, or all of them.

NUTS, LEGUMES, CEREALS

FRUITS & VEGETABLES

MEAT ALTERNATIVES

WHAT DOES THE PORTFOLIO DIET LOOK LIKE?

Expected LDL-Cholesterol lowering:

1 NUTS 45g DAILY

All nuts are good for your heart and cholesterol and contrary to concerns do not contribute to weight gain. Add nuts as a snack between meals, adding to salads, cereals, or yogurt. Trying nut butter on your toast is an option. 45g is about a handful of nuts. If allergic to peanuts or tree nuts, try seeds.

MIXED NUTS ALMONDS PEANUTS

NUT BUTTERS PISTACHIOS WALNUTS

5 - 10%

2 PLANT PROTEIN 50g DAILY

This is the most challenging component of the Portfolio diet. Start by trying to get 25g daily. Consider replacing milk with soy milk, try tofu, soy nuts and beans. .

CHICKPEAS PEAS TEMPEH VEGGIE BURGER TOFU SOY BEANS

LENTILS BEANS VEGGIE DOG SOY MILK SOY DELI SLICES

5 - 10%

3 VISCOUS (STICKY) FIBRE 20g DAILY

Aim to eat 2 servings of oatmeal, beans, lentils, and chickpeas a day. Replace bread with rye or pumpernickel or oatcakes. Eat at least 5 servings of fruit and vegetables every day. Aim to eat 2 servings per day of oatmeal, barley, or cereals enriched with psyllium or oat bran. Replace white bread with whole grain oat breads. Put oat bran or psyllium into smoothies. Eat at least 5 servings per day of vegetables (eggplant, okra) and fruit (apples, oranges, berries) high in viscous fibre.

APPLE OKRA EGGPLANT

PSYLLIUM STRAWBERRIES OATMEAL

OAT BRAN CEREAL BARLEY

5 - 10%

4 PLANT STEROLS 2g DAILY

These occur naturally (soyabean, corn, squash, etc.) but to get this amount of sterol you will require fortified foods such as spreads, juices, yogurt, milk and even supplements as part of a meal.

PLANT STEROL MARGARINE OILS JUICES YOGURT

PLANT STEROL FORTIFIED

5 - 10%

TOTAL: ~30%

Statins, the most effective class of cholesterol-lowering medications, reduce cholesterol by 20-60%

IT'S NOT ABOUT ONE BIG CHANGE. IT'S NOT ALL OR NOTHING. JUST START BY INTRODUCING ONE COMPONENT TO YOUR DIET AND BUILD FROM THERE.

THIS PORTFOLIO DIET IS FOR "REAL PEOPLE IN THE REAL WORLD"

— DR. DAVID JENKINS, CREATOR OF THE PORTFOLIO DIET

David JA Jenkins MD, PhD, DSc, Cyril WC Kendall PhD, Lilisha Burris MHSc, RD, John L Sievenpiper MD, PhD, FRCPC, Michael F. Evans MD, CCFP, Emily Nicholas Angl BSc

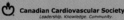

Canadian Cardiovascular Society
Leadership. Knowledge. Community.

The Joannah & Brian Lawson Centre for Child Nutrition
UNIVERSITY OF TORONTO

St. Michael's
Inspired Care.
Inspiring Science.

REFERENCES

[1] Jenkins D.J.A., Kendall C.W.C., Faulkner D.A., Kemp T., Marchie A., Nguyen T.H., et al. Long-term effects of a plant-based dietary portfolio of cholesterol-lowering foods on blood pressure. Eur J Clin Nutr 2008;62(6):781–8. https://doi.org/10.1038/sj.ejcn.1602768.

[2] Jenkins D.J.A., Jones P.J.H., Lamarche B., Kendall C.W.C., Faulkner D., Cermakova L., et al. Effect of a dietary portfolio of cholesterol-lowering foods given at 2 levels of intensity of dietary advice on serum lipids in hyperlipidemia: a randomized controlled trial. J Am Med Assoc 2011;306(8):831–9. https://doi.org/10.1001/jama.2011.1202.

[3] Jenkins D.J.A., Kendall C.W.C., Marchie A., Faulkner D.A., Wong J.M.W., de Souza R., et al. Direct comparison of a dietary portfolio of cholesterol-lowering foods with a statin in hypercholesterolemic participants. Am J Clin Nutr 2005;81(2):380–7.

[4] Anderson T.J., Grégoire J., Pearson G.J., Barry A.R., Couture P., Dawes M., et al. 2016 Canadian Cardiovascular Society Guidelines for the management of dyslipidemia for the prevention of cardiovascular disease in the adult. Can J Cardiol 2016;32(11):1263–82. https://doi.org/10.1016/j.cjca.2016.07.510.

[5] Catapano A.L., Graham I., De Backer G., Wiklund O., Chapman M.J., Drexel H., et al. 2016 ESC/EAS guidelines for the management of dyslipidaemias. Rev Esp Cardiol 2017;70(2):115. https://doi.org/10.1016/j.rec.2017.01.002.

[6] Heart U.K. The Portfolio Diet. Retrieved from https://heartuk.org.uk/files/uploads/documents/huk_fs_d01_theportfoliodiet.pdf; 2006.

[7] Gigleux I., Jenkins D.J.A., Kendall C.W.C., Marchie A., Faulkner D.A., Wong J.M.W., et al. Comparison of a dietary portfolio diet of cholesterol-lowering foods and a statin on LDL particle size phenotype in hypercholesterolaemic participants. Br J Nutr 2007;98(6):1229–36. https://doi.org/10.1017/S0007114507781461.

[8] Lamarche B., Desroches S., Jenkins D.J.A., Kendall C.W.C., Marchie A., Faulkner D., et al. Combined effects of a dietary portfolio of plant sterols, vegetable protein, viscous fibre and almonds on LDL particle size. Br J Nutr 2004;92(4):657–63.

[9] Lopez-Garcia E., Schulze M.B., Meigs J.B., Manson J.E., Rifai N., Stampfer M.J., et al. Consumption of trans fatty acids is related to plasma biomarkers of inflammation and endothelial dysfunction. J Nutr 2005;135(3):562–6. https://doi.org/10.1093/jn/135.3.562.

[10] Lu H., Ouyang W., Huang C. Inflammation, a key event in cancer development. Mol Cancer Res 2006;4(4):221–33. https://doi.org/10.1158/1541-7786.MCR-05-0261.

[11] Jenkins D.J.A., Jones P.J., Frohlich J., Lamarche B., Ireland C., Nishi S.K., et al. The effect of a dietary portfolio compared to a DASH-type diet on blood pressure. Nutr Metab Cardiovasc Dis 2015;25(12):1132–9. https://doi.org/10.1016/j.numecd.2015.08.006.

[12] Salas-Salvado J., Bullo M., Babio N., Martinez-Gonzalez M.A., Ibarrola-Jurado N., Basora J., et al. Reduction in the incidence of type 2 diabetes with the Mediterranean diet: results of the PREDIMED-Reus nutrition intervention randomized trial. Diabetes Care 2011;34(1):14–9. https://doi.org/10.2337/dc10-1288.

Chapter 1

The Power of Diet in CVD Risk Factor Reduction

INTRODUCTION

Cardiovascular disease (CVD) is the number one cause of death globally. The term CVD covers a range of conditions affecting the heart, but is primarily used to describe conditions that result in narrowed or blocked blood vessels, which can lead to heart attack, stroke, and angina (chronic chest pain). Many forms of heart disease are largely preventable and improved with diet and lifestyle changes. As a result, a number of evidence-based diets have been designed to target CVD. One such diet is The Dietary Portfolio, others include: The Mediterranean diet, the Okinawan diet, the Nordic diet, a low glycemic index (GI) diet, the Dietary Approach to Stop Hypertension (DASH) diet, Atkins, and Eco-Atkins. Most of these diets were developed to mimic the traditional diets of populations known to have low levels of CVD risk factors and incidence of CVD. In contrast, the Portfolio diet deconstructed this approach, combining food components known to individually reduce the risk of CVD and situating them in the context of a healthy eating pattern. By combining foods known to lower CVD risk factors, their effect, on lowering blood lipids, was maximized. As the Portfolio diet includes key components from other diets, there are similarities in terms of its effects on CVD risk factors with other evidence-based diets. This chapter will discuss the similarities and differences between evidence-based diets and how these relate to their effects on three CVD risk factors: body weight, blood pressure, and blood lipids. A summary of the following discussion on risk factors addressed by evidence based diets can be found in Table 1.1. This chapter will also give an overview of each these risk factors for those wishing to gain a more thorough understanding of their importance and relevance.

Cardiovascular disease (CVD) includes; coronary heart disease (CHD), peripheral artery disease, and stroke.

Key points:
Weight
- Low-carbohydrate diets like Atkins may negatively affect CVD risk in the long run once the benefits of weight-loss have plateaued as low-density lipoprotein cholesterol (LDL-C) levels remain high.
  - Eco-Atkins, a high protein, plant-based variation of the Atkins diet, resulted in reductions in both waistlines and LDL-C.
- Both a low GI diet and the Portfolio diet have been observed to reduce body weight over the long term.

The Portfolio Diet for Cardiovascular Disease Risk Reduction. https://doi.org/10.1016/B978-0-12-810510-8.00001-7

TABLE 1.1 Comparison of Dietary Patterns and Main Components

Diet	CVD-Related Target	Key Components of Dietary Pattern
Low Carb diet	WL	Atkins [1] – Carbohydrate <20% of total energy intake, high in animal products, high in fat. Eco-Atkins [2] – Increased protein and fat from gluten and soy products, nuts, and vegetable oils.
DASH [3]	BP	– High in low fat fermented dairy products, vegetables, whole grains, poultry, fish, and nuts. – Low consumption of red meat recommended. Omni heart variations [4]: – 10% of carbohydrate derived energy replaced with mostly plant-based sources of protein. – 10% of carbohydrate derived energy replaced with unsaturated fats (mainly monounsaturated).
Low Glycemic Index Diet [5]	BL, WL	– No specified portion sizes or optimal calories specified. Relies on consumer constructed plan based on GI/GL values of foods. – Typically high in plant-based foods and whole grains. – The Zone, Sugar Buster and the Slow Carb Diet are all based off this diet and contain more specific parameters.
Mediterranean [6,7]	Overall CVD	– High in fresh fruits, vegetables, whole grains, fatty fish, nuts, plant-based oils, and margarine. – Low in red meat.
Nordic [8]	Overall CVD	– Including oily fish (salmon and mackerel), vegetables, roots, legumes, fruits, berries, and wholegrain cereals (oat, rye, and barley).
Okinawan [9]	BP	– Traditional diet rich in fish, seaweed, soybean products, vegetables, and green tea.
Ornish [10]	Overall CVD	– Vegetarian, high fiber and low fat diet.
Portfolio Diet [11]	BL	– High in whole grains, legumes, plant sterols (plant based oils), nuts, viscous fiber foods (barley, okra, psyllium), berries and soy.
Pritikin [12]	Overall CVD	– Low fat, emphasis on complex carbohydrates: less than 10% of calories from fat, 10%–15% from protein and 75%–80% complex carbohydrates. – Added exercise component.
Vegetarian [13]	BP, BL	– High in fruits and vegetables, legumes, nuts, whole grains, moderate amounts of dairy and eggs.
Zone [14]	GH	– Based on macronutrient distribution, with heavy emphasis on protein. Each daily meal consists of: 30% protein, 40% carbohydrates, 30% fats.

WL, weight loss; *BP,* blood pressure; *CVD,* cardio vascular disease; *BL,* blood lipids; *GH,* general health; *NA,* not applicable; *GI,* glycemic index; *GL,* glycemic load.

Blood pressure
- Significant reductions in blood pressure (BP) were observed for the DASH diet, the Vegetarian diet, the Portfolio diet, and the Mediterranean diet, likely due to the high consumption of fruits and vegetables common to these diets.

Blood lipids
- LDL-C reductions are seen in the Portfolio, Nordic, Mediterranean, DASH, and Low GI diet.
- There is strong evidence for the benefits of reducing LDL-C irrespective of baseline LDL-C levels.
- LDL-C: HDL-C (low-density lipoprotein: high-density lipoprotein) ratios may be more predictive of CVD than either LDL-C or HDL-C (high-density lipoprotein) alone.
  - The Portfolio and Mediterranean Diets both decreased the ratio of LDL-C:HDL-C, likely because of the high concentration of monounsaturated fats from plant-based oils and nuts found in both diets.

Framingham equation:
  - Reduced 10 year risk of CHD/CVD found in Portfolio diet, vegetarian diets and the DASH diet.

A HEALTHY WEIGHT

How Important Is It?

Body weight has been shown to have strong associations with CVD [15,16]. One of the most common ways scientists report body weight is using an individual's body mass index (BMI). BMI uses a formula that considers an individual's height and body weight with desirable BMI being equal to or greater than 25 kg/m^2. This calculation helps standardize the natural differences you see in body weight for a short person compared to a tall person. See the bubble on the left to learn how to calculate your BMI at home. Compare this number to the risk categories shown in Box 1.1 to see whether your body weight puts you at risk for CVD. For example, if your weight was 70 kg and your height was 1.64 m, then your BMI would be 26.0 kg/m^2 putting you in a body weight category associated with lower risk. However, BMI and its association with risk status should be interpreted with caution as certain factors may influence the applicability of BMI as an indicator of CVD risk. For example, an individual with high muscle mass may have a high BMI, but is not actually "at risk" because the weight of muscle makes them fall into a higher BMI range. This individual may have a much lower risk of CVD compared to an individual with the same BMI but lower muscle mass. Also ethnicity plays a role. For individuals of Asian decent, lower BMI's may be more desirable (less than 25 kg/m^2) [20], while individuals of African descent may have higher BMI's (greater than 25 kg/m^2), without being at greater risk for chronic disease [21–23].

Other factors influencing the association between BMI and CVD include: fat distribution, history of weight gain, cholesterol, and hypertension. Measurements of waist circumference, particularly for higher BMI's (25–35 kg/m^2) are now increasingly used as a tool in determining the risk of CVD. Unlike BMI, waist circumference takes into consideration the fat distribution. This method provides a way to determine

The bubble on the left reads:

BMI can be determined at home. Simply plug your information into this formula:

body weight (kg) ÷ height (m^2)

See risk categories in Box 1.1 to see where you stand.

BOX 1.1 BMI and CVD Risk for Adults Over 20 Years of Age[1] [17]

- A BMI of 18.5–24.9 kg/m^2 is associated with little to no increased risk.
- A BMI of 25–29.9 kg/m^2 is associated with low risk.
- A BMI of 30–34.9 kg/m^2 is associated with moderate risk.
- A BMI of 35–39.9 kg/m^2 is associated with high risk.
- A BMI above 40 kg/m^2 is associated with very high risk.

  These risk categories apply unless visceral fat is high, or the individual has gained over 10 kg since the age of 18. In that case the risks are substantially higher [18].

  Waist circumference and risk:
- Greater than 40 in (102 cm) in men = increased risk.
- Greater than 35 in (88 cm) in women = increased risk.
- If BMI above 35, waist circumference does not factor in as individuals are already at high risk [19].

---

[1] These may be different for different populations.

---

levels of visceral fat (fat around the abdomen), which is known to increase CVD risk to a greater extent than other fat distribution patterns [19]. Guidelines on interpreting waist circumference can be found in Box 1.1.

Due to the complex relationship between BMI and CVD, weight-loss may be much more important for some individuals than others. As of yet there is no quick fix or pharmaceutical agent that can rival the impact of proper diet and exercise in maintaining and achieving a healthy weight. If you are still unsure if weight-loss is appropriate for you, consult your physician or a dietitian.

Weight-Loss Targeted Diets

While caloric restriction has been shown to have the greatest influence on weight-loss, it is equally important to take into account the quality of food in the diet [24,25]. Simply cutting calories may be an effective strategy for losing weight in the short-term but is often unsustainable. Additionally, this approach does not address other risk factors of CVD. Following an evidence-based diet is the best way to achieve a healthy weight while reducing other factors that may contribute to CVD. Low carbohydrate diets such as the Atkins diet are centered around intake of high protein and/or fatty foods. Unlike the other diets examined in this section the main mechanism by which low-carbohydrate diets are thought to reduce Coronary Heart Disease (CHD), a subtype of CVD, is through weight-loss. A recent meta-analysis examining popular weight-loss diets without advocating caloric restriction found that Atkins had greatest amount of evidence supporting both short-term and long-term reductions in body weight [26]. However, high animal product content, saturated fat, and dietary cholesterol put this diet at odds with what we know about effective CVD prevention. A meta-analysis by Mansoor et al. [27] analyzed randomized

> Low carbohydrate diets need not be meat based, in the Eco-Atkins diet, the protein and fat come exclusively from vegetable sources.

controlled trials of Atkins diets and found that while these studies did indeed show significant weight-loss, total cholesterol also increased. High cholesterol levels, particularly high LDL-C levels, are major risk factors for CHD. In this way the Atkins diet creates a trade-off scenario where both weight-loss and reductions in total cholesterol cannot be simultaneously achieved. This conflict has led to the conclusion that Atkins and low-carbohydrate diets may not be suitable for individuals at risk for CVD, particularly in the long-run once weight-loss has plateaued [27]. One possible exception to this finding is the Eco-Atkins diet which is also a low-carbohydrate diet but is entirely plant-based and centered around Portfolio diet foods [28]. The Eco-Atkins diet was aptly named as it follows the same low-carbohydrate high-protein and high-fat principle as the original Atkins diet but with reduced ecological impact relating to the lack of animal products. Eco-Atkins was able to achieve significant weight-loss similar to that of other "Atkins like" diets while still reducing LDL-C and total cholesterol. As seen in Table 1.1, Portfolio foods were incorporated into the Eco-Atkins diet. The cholesterol lowering effects of Portfolio diet foods have been proposed to account for the reduction in LDL-C seen in Eco-Atkins. If weight-loss is your goal, then the Eco-Atkins diet which includes Portfolio components may be beneficial. While participants in many of the Portfolio diet studies did not lose weight this effect was a result of being expressly asked not to, so the effects of the food components could be evaluated independent of weight-loss. In a later, longer, and multicenter trial where such instructions were not given, a modest, but significant amount of weight-loss was observed (around 1.5 kg) [29].

Low GI foods are thought to be more satiating or filling as blood glucose levels tend not to follow the "peak and crash pattern" seen with high GI foods and instead provide a constant supply of energy. This satiating effect has been proposed to aid in weight-loss. Some support for this hypothesis was seen in a study which compared the effect of a vegan/plant-based, low GI diet with an American Diabetes Association diet targeting weight-loss in individuals with type 2 diabetes [30]. This study demonstrated that the GI of the diet significantly predicted weight change irrespective of the actual background diet and for every 1-unit decrease in GI, about 0.2 kg was lost [30]. In a 6-month study in which 210 individuals with type 2 diabetes were randomized to follow either a low GI diet or a diet high in cereal fiber, participants on the low GI arm lost around 3 kg, while those on the high cereal arm lost around 1.5 kg [31]. This result has led to the conclusion that a low GI diet may contribute to modest weight-loss. The Portfolio diet uses plant-based, low GI, high fiber foods similar to those consumed in the study discussed above. Depending of the preference and goals of the the individual the Portfolio diet can contain more or less of these foods. See Box 1.2 below for tips on how to incorporate low GI foods into the Dietary Portfolio.

Other notable findings include weight-loss reductions seen in overweight and obese individuals on the Mediterranean diet. One study was able to achieve an 8.9 kg reduction in 12 months in obese individuals

---

BOX 1.2 The Glycemic Index and the Dietary Portfolio

A low glycemic index (GI) diet can aid in the management and prevention of type 2 diabetes [31–33]. The Dietary Portfolio can be easily adapted to include foods with a low GI, further lowering the GI of the diet. Websites such as glycemicindex.com run by the University of Sydney provide useful tools for calculating the GI and glycemic load (GL) of foods [34]. As a rough guide, a GI of 55 or less is considered low for a carbohydrate food. GL can be useful as it takes into account the total glycemic impact of a meal.

with type 2 diabetes on a low-carb Mediterranean diet compared to 7.4 kg on a traditional Mediterranean diet [35]. Another study looking at obese and overweight individuals without type 2 diabetes found less substantial but significant reductions of 4.2 kg over the course of 12 months [36]. The Ornish diet was able to achieve reductions of 2.2 kg after 12 months in premenopausal obese and overweight women [37]. Several trials looking at the Zone diet found weight-loss ranging from 1.5–3.2 kg after 12 months [37]. Among the individuals with an increased waist circumference the Nordic diet saw an average decrease in body weight of 4.7 kg [38]. A meta-analysis on vegetarian eating patterns over the median of 18 weeks found significant reductions of 2.02 kg in those on a vegetarian diet and somewhat greater reductions of 2.52 kg for those on a vegan diet [39]. Like the Portfolio diet, all diets discussed in this paragraph have relatively low animal product consumption compared to the average western diet and place strong emphasis on fruit and vegetable consumption.

As illustrated by the studies discussed above there are many paths to weight-loss. While some of these studies demonstrated greater weight-loss effects than others, direct comparison of results should be made with caution as many of the study criteria and the starting points of the participants will make them more or less likely to lose weight. What we can conclude is that there are many effective evidence based approaches to weight-loss. The Dietary Portfolio has many factors in common with diets which have been shown to promote weight-loss. Additionally, there is direct evidence to show that the Portfolio diet can help to maintain a healthy body weight or achieve decreases in weight when consumed in the context of diets such as the Eco-Atkins.

BLOOD PRESSURE

How High Is Too High

Hypertension or high blood pressure (BP) is a known risk factor for stroke and mortality from CVD. A large cohort study including over 1.25 million patients found that those with baseline hypertension had a ~63% risk of CVD compared with ~46% risk of CVD for those with normal BP [40]. What constitutes hypertensive BP is up for some debate as the "white coat effect" (see bubble to the left) and variation throughout the day can factor in substantially [41]. The diagnostic criteria suggested by American College of Cardiology (ACC) and American Heart Association (AHA) can be found in Box 1.3 [44].

The "white coat effect" refers to unusually high blood pressure readings seen in a medical setting due to tension and stress, i.e. where health professionals wear white coats.

Like weight-loss many factors influence the likelihood and extent of hypertension in an individual. Individuals of African or Hispanic descent have been found to have the highest risk of hypertension. Individuals of African descent who experience hypertension tend to have increased severity of symptoms due to its association with greater organ damage in this population [45]. Age also factors into diagnosis. For patients under the age of 50 diastolic BP is the most predictive of CVD compared to those over age of 60 where total pulse pressure (the force with which the heart contracts) was the major indicator (see Box 1.3 for how to calculate pulse pressure) [46]. Genetics also play a strong role. Individuals who have hypertensive parents are twice as likely to develop hypertension themselves [47]. Studies looking at trends in a population over time, have shown that genetics account for around 30% of the risk for developing hypertension [47].

BOX 1.3 Blood Pressure Readings (in mmHg)

Blood Pressure [42]
Normal:
<120 SBP and 80 DBP
Elevated:
120–129 SBP and <80 DBP
Hypertension:
- Stage 1: 130–139 SBP or 80–89 DBP.
- Stage 2: ≥140 SBP or ≥90 DBP.

Pulse Pressure [43]
Calculation:
SPB-DBP
For example, for a BP of 120/80, pulse pressure would be 40.
At Risk:
- <40
- >60

SBP, systolic BP; *DBP*, diastolic BP.

Blood Pressure Lowering Diets

The DASH diet was developed with the specific intent of lowering blood pressure. The diet was formulated using the observation that individuals following vegetarian dietary patterns tended to have lower BP than those including meat in their diet. While DASH is not a fully vegetarian diet, it advocates limiting meat consumption and includes generous servings of fruit, vegetables, and low-fat dairy products. Many studies have evaluated the DASH diet and it has subsequently been refined and advanced over time. DASH has shown the ability to reduce systolic blood pressure (SBP) by 5.5 mmHg and diastolic blood pressure (DBP) by 3 mmHg [3]. In individuals with hypertension the effects were even greater, with a mean reduction of 11.4 mmHg in SBP and 5.5 mmHg in DBP in studies where all the food was provided and prepared. This result demonstrates that large reductions in BP can be achieved if the diet is strictly followed. Further advancements were made in the "optimal macro nutrient intake" or OMNI-heart trial showing that a high protein and low saturated fat diet was the most beneficial in hypertensive individuals [48].

The DASH-sodium study looked at the effect of the DASH diet with different levels of sodium intake and determined that reduction in sodium was associated with further improvement in BP [49]. These results were found in all subgroups including hypertensive and individuals of African descent in which the greatest effects were observed. Interestingly, the Portfolio diet did not find such a correlation between salt intake with BP. Another seeming contradiction was seen in the "Japanese" or Okinawa diet. This diet is modeled on the traditional diet of the population of Okinawa, one of the longest-lived populations in the world. They consume high amounts of fish, soy, seaweed, green tea, vegetables, and salt. Despite the prevalence of high salt consumption and hypertension in this population, CVD incidence remains low [9]. Serious salt restriction for the prevention of CVD is a debated concept in the scientific community (see Box 1.4 for more details on the salt debate).

When the Portfolio diet and a DASH-type control diet were compared head-to-head, the Portfolio diet reduced BP to a greater extent compared to the DASH-type diet (2.1 mmHg compared to 1.8 mmHg, respectively) [54]. Nuts, soy, and viscous fiber were all positively linked to these reductions. Results from this study should be interpreted cautiously as significance was only reached toward the end of the study at the week 24 mark. It is possible that a longer time frame would have been needed to properly observe the results. Nevertheless, the BP lowering aspect of the Portfolio diet shows promise but warrants further exploration.

BOX 1.4 Salt of the Earth

The last major review on sodium by the National Academy of Science [formally the Institute of Medicine (IOM)] was conducted in 2013 [50] and updated in 2019 [51]. Both reports found that higher levels of sodium consumption and CVD were related, and conclude that sodium levels should be reduced to below 2300 mg per day (about 1 tsp.). Not all evidence has supported the conclusion that salt increases CVD risk. Song et al. [52], found that in individuals at low risk of heart failure (NYHA I/II), consuming <2 g/day had a higher risk for hospitalization or death compared to those consuming 2–3 g/day. A 2011 Cochrane review found no conclusive data concerning salt and its role in CVD, concluding that further investigation was necessary [53]. As salt can make healthy foods more palatable, these results pose the question of whether or not foods with higher salt content or added salt, may be permitted in the diet if they increase the total amount of healthy foods consumed as a result.

Although not a primary outcome of the Mediterranean diet, blood pressure reductions have been linked to this dietary pattern [55]. These reductions are moderate but significant (0.65 mmHg reduction in DBP) and possibly related to the monounsaturated fat content from the 30 g/day of mixed nuts the Mediterranean diet recommends participants to consume. This intake is similar to the mean nut intake of the Portfolio diet of 36 g/day.

Other notable reductions in BP were found in a meta-analysis of low GI diets, which observed a decrease in SBP:DBP by 1.1 mmHg per 10 unit reduction of GI in healthy individuals [56]. In a randomized control trial, the Nordic diet showed significant reductions in BP [57]. This reduction has been proposed to be the result of the large quantities of blueberries recommended in this dietary pattern. Berries are sources of potassium very high in polyphenols, particularly flavonoids which have been shown to reduce cholesterol in randomized control trials [58]. Similar to the Nordic diet, the Portfolio diet also includes a recommendation for berries, potentially adding to the blood pressure lowering effect. Other components of the Portfolio diet like soy have also been shown to independently lower blood pressure in a number of meta-analyses [59], as has viscous fiber [60].

BLOOD LIPIDS

Good vs. Bad

Reduction in LDL-C levels has been the primary outcome of the Portfolio studies. High levels of LDL-C sometimes referred to as the "bad cholesterol" has been linked by numerous studies to the development of CHD and CVD [61–68]. This association is because LDL-C promotes the formation of cholesterol plaques in arterial walls. This plaque can block blood flow to the heart or brain so it is important to keep LDL-C levels low. Interestingly, reducing LDL-C reduces CVD risk irrespective of LDL-C baseline level, that is, the health benefits of reducing LDL-C levels are applicable to the whole population, even healthy adults with normal LDL-C levels.

Lowering of LDL-C levels reduces CVD events irrespective of your starting LDL-C level.

While reduction in LDL-C levels has been the primary aim of both drug and dietary interventions, low HDL-C levels, sometimes referred to as the "good cholesterol," are also associated with increased CVD risk. This increased risk may be due to the ability of HDL-C to remove LDL-C from plaque in the arterial wall [69]. HDL-C brings this cholesterol back to the liver to be removed. Lifestyle modifications such as exercise, healthy body weight, cessation of smoking, low to moderate alcohol consumption, may increase HDL-C levels by as much as 10%–13% [70,71]. As both LDL-C and HDL-C are independent predictors of CVD risk, it is not surprising that the ratio is an even better predictor than either LDL-C or HDL-C alone. It has been suggested that an even stronger predictor of CVD risk is the ratio of apolipoprotein B, (a major component of LDL-C) to apolipoprotein A1 (a major component of HDL-C) (ApoB:ApoA1) [72]. In the original Portfolio diet study, the ratio of ApoB:ApoA1 was measured and found to decrease between the start and the finish of the trial.

Caucasians are more likely to have hyperlipidemia [74].

However, despite evidence that this ratio may be a stronger predictor, it is not as commonly used as the ratio of LDL-C:HDL-C. When getting blood lipids checked typically only LDL-C levels are measured with ideal levels falling below 1.8 mmol/L or 70 mg/dl (US). For more information, on ApoB:ApoA1 and LDL-C:HDL-C levels refer to the Canadian Cardiovascular Society Guidelines on dyslipidemia [73].

The importance of blood lipid levels is dependent on a number of criteria. If you have not previously been tested, see Box 1.5 "When to Get Checked" to determine whether or not you should be screened by your physician. In addition to the factors discussed in Box 1.5, race, ethnicity, and gender also factor into the risk of developing high blood cholesterol (hypercholesterolemia) and high blood lipids (hyperlipidemia). Unlike blood pressure, Caucasian groups are the most likely to develop these disorders than other ethnic groups [74].

BOX 1.5 When to Get Checked [73]

- If you are over the age of 40.
- If you fall into any of the following categories or have any of the following conditions:
  - Some form of CVD.
  - Diabetes mellitus.
  - Smoke cigarettes.
  - Family history of CVD.
  - Family history of dyslipidemia.
  - Obesity.
  - Inflammatory diseases.
  - HIV.
  - Erectile dysfunction.

Cholesterol Lowering Diets

Statin trials have demonstrated large decreases in LDL-C and consequent reductions in all-cause mortality and myocardial events. The efficacy of statins was one of the reasons that a head-to-head comparison between the Portfolio diet and the statin drug Lovastatin was conducted. In a 4-week head-to-head Portfolio vs. Lovastatin diet study, the reduction in LDL-C was 33% after Lovastatin and 30% after the Portfolio diet. To put this in perspective, in the Air/Force/Texas Coronary Atherosclerosis Prevention Study, Lovastatin reduced LDL-C levels by 25% and reduced the incidence of a first cardiac event in a low-risk population (the average LDL-C level was 3.9 mmol/L) [61]. The Portfolio vs. Lovastatin trail was one of the first to demonstrate that dietary interventions could be as potent as the first-generation statins [75].

The early Portfolio diet did not show increases in HDL-C, but after the addition of monounsaturated fatty acids (MUFA) in the form of high oleic sunflower oil to the existing Portfolio diet foods, HDL-C levels increased [76] and an even greater decrease in the LDL-C:HDL-C ratio was observed. Similarly, the Mediterranean diet which is also high in MUFA from olive oil and nuts has shown similar reductions in LDL-C:HDL-C ratios.

Like the Portfolio diet, many other dietary patterns have been investigated to determine their ability to lower total cholesterol, LDL-C, and the LDL-C:HDL-C ratio. The Nordic diet, like the Portfolio diet, which emphasizes whole grains and viscous fibers such as oats and barley, saw a 21% reduction in LDL-C. Although promising, these results should be interpreted with caution as the study was relatively short in duration, lasting 6 weeks [77]. A meta-analysis of vegetarian diets which included protein from sources such as nuts and soy showed significant reductions in total cholesterol [13]. The Mediterranean diet, which includes large quantities of plant sterols and nuts, found significant reductions in LDL-C versus a low-fat control diet [55]. The DASH diet, which focuses on increasing fruit, vegetable, and fiber intake found significant reductions in total cholesterol and LDL-C relative to the control: a typical American diet which was lower in fruits, vegetables and fiber while being higher in sweets [78]. Low GI diets have been shown to raise HDL-C in type 2 diabetic patients compared to a high fiber cereal control [31,79]. The starchy foods included in the Portfolio diet are also low GI and can be further adapted to a lower GI diet using tools mentioned in Box 1.2.

> The Portfolio diet can achieve similar LDL-C reductions as seen with first-generation statins.

> Both HDL-C and apolipoprotein A1 levels increase when consuming a high MUFA Portfolio diet.

Although many diets targeting cholesterol have features in common with the Portfolio diet, there are some diets which, despite being very different, have shown modest reductions in LDL-C. When the Atkins, the Ornish, and the Zone diet were tested in a head-to-head comparison, all diets showed around a 10% reduction in the LDL-C:HDL-C ratio after 1 year, with no significant advantage being conferred by any diet [80]. A recent study investigating the effects of a short-term Pritikin diet and exercise program, found 10%–15% LDL-C reductions in patients with metabolic syndrome [81]. As this study was 10–15 days in length, these results must be interpreted with caution.

Considering these studies collectively, it is clear that there are many dietary interventions that can reduce cholesterol. Many of the diets which have had success have characteristics in common with the Portfolio diet including the incorporation of viscous fiber, fruits, vegetables, soy products, plant sterols, and nuts.

THE FRAMINGHAM EQUATION

The Framingham equation is one of the ways to measure CVD risk [82,83]. It uses a combination of the factors discussed earlier (blood pressure, body weight, cholesterol) along with other factors such as patient history to predict the risk of an individual developing cardiovascular disease over the next 10 years.

Originally, the Framingham equation only assessed risk for developing CHD, a specific subtype of CVD. Over time as factors were added the equation, its predictive ability was expanded to include CVD risk. This equation is one of the many tools used by physicians to decide whether pharmacotherapy and/or lifestyle changes are recommended to reduce the risk of CVD. Certain biomarkers are especially predictive of the risk of cardiac event once it has been determined that individuals have an elevated risk of CVD. C-reactive protein is one such biomarker, see Box 1.6 on C-reactive protein for more details.

BOX 1.6 C-Reactive Protein

C-reactive protein (CRP) is a marker of inflammation and can contribute to the narrowing of the arteries which can result in a heart attack. It is an acute phase protein which means that it reflects changes in the body in real-time. Men with elevated CRP levels have been found have three times the risk of having a heart attack or stroke than those with low levels [84]. In the Harvard Women's Health Study, CRP has been found to be more predictive of heart attack and stroke than high cholesterol levels [85]. For those who already have an elevated risk of developing CVD, CRP is an especially predictive marker for risk of heart attack [86].

The American Heart Association defines risk categories as follows [86]:

- Low risk. hs-CRP level < 2.0 (mg/L).
- High risk. hs-CRP level > 2.0 (mg/L).

However, it is important to note that many factors can contribute to increased CRP levels. For example, a common cold can increase inflammation and thus CRP as a consequence. Therefore, this measure should not be used on its own to predict risk of heart attack.

THE FRAMINGHAM EQUATION AND DIET

The precise criteria for this calculation are available on the Canadian Cardiovascular Society website.[1]

In the original Portfolio diet study [11], the Framingham equation was used to illustrate the effectiveness of diet on prevention of future CHD risk. Those consuming the Portfolio diet were found to decrease their 10-year risk of developing CHD by 25% compared to the control. Others studies have also used the Framingham equation to predict CVD/CHD risk. Those consuming a vegetarian diet were found to have a lower Framingham risk score compared to those consuming an omnivorous diet [87]. The DASH diet has also been found to reduce CVD risk compared to the control [88]. The Framingham equation is just one tool used to predict CVD risk and evaluate the effectiveness of a diet or treatment. Every person will have their own set of additional risk such as ethnicity and family history which are not fully taken into account by models such as the Framingham equation. However it is still a useful tool in determining an estimate of risk, combining the risk factors discussed above.

Different diets have different strengths. By targeting specific risk factors, diets can complement drug-based interventions, act as a nondrug-based alternative or work as a preventive strategy for CVD.

Targeted approaches also offer guidance to individuals with specific health concerns. While resources like a countries' dietary guidelines are useful to the general population, certain individuals may have different needs because of genetics or lifestyle. Targeted diets allow individuals to address their specific health concerns. This ability makes them an important tool in the arsenal of CVD prevention. The Dietary Portfolio can accommodate many lifestyles and dietary preferences. While it is effective at reducing all risk factors discussed above, its greatest strength lies in its ability to alter blood lipids. In the Portfolio studies

---

[1] https://www.ccs.ca/images/Guidelines/Tools_and_Calculators_En/FRS_eng_2017_fnl1.pdf.

this capacity has, in part, lead to the strong reductions in 10 year risk predicted by the Framingham equation. The next chapter further investigates what the Dietary Portfolio is and the evidence base behind it.

REFERENCES

[1] Atkins R. Dr. Atkins' new diet revolution. Avon Books; 1998.

[2] Jenkins DJA, et al. The effect of a plant-based low-carbohydrate ('Eco-Atkins') diet on body weight and blood lipid concentrations in hyperlipidemic subjects. Arch Intern Med 2009;169:1046–54.

[3] Appel LJ, Moore TJ, Obarzanek E, Vollmer WM, Svetkey LP, Sacks FM, et al. A clinical trial of the effects of dietary patterns on blood pressure. N Engl J Med 1997;336(16):1117–24. https://doi.org/10.1056/NEJM199704173361601.

[4] Appel LJ, et al. Effects of protein, monounsaturated fat, and carbohydrate intake on blood pressure and serum lipids. JAMA 2005;294:2455.

[5] Jenkins D, et al. Glycemic index of foods: a physiological basis for carbohydrate exchange. Am J Clin Nutr 1981;34:362–6.

[6] Fidanza F, Puddu V, Imbimbo AB, Menotti A, Keys A. Coronary heart disease in seven countries. VII Five-year experience in rural Italy. Circulation 1970;41:163–75.

[7] Aravanis C, Corcondilas A, Dontas AS, Lekos D, Keys A. Coronary heart disease in seven countries. IX The Greek islands of Crete and Corfu. Circulation 1970;41:I88–100.

[8] Adamsson V, et al. What is a healthy Nordic diet? Foods and nutrients in the NORD1ET study. Food Nutr Res 2012;56:18189.

[9] Shimazu T, Kuriyama S, Hozawa A, Ohmori K, Sato Y, Nakaya N, et al. Dietary patterns and cardiovascular disease mortality in Japan: a prospective cohort study. Int J Epidemiol 2007;36(3):600–9. https://doi.org/10.1093/ije/dym005.

[10] Ornish DD. Dr. Dean Ornish's program for reversing heart disease. New York: Random House Publishing Group; 2010.

[11] Jenkins DJA, Kendall CWC, Marchie A, Faulkner DA, Wong JMW, de Souza R, et al. Effects of a dietary portfolio of cholesterol-lowering foods vs lovastatin on serum lipids and C-reactive protein. JAMA 2003;290(4):502–10. https://doi.org/10.1001/jama.290.4.502.

[12] Pritikin N, McGrady PM. The Pritikin program for diet and exercise. East Melbourne, VIC: Schwartz Publishing; 1980.

[13] Wang F, Zheng J, Yang B, Jiang J, Fu Y, Li D. Effects of vegetarian diets on blood lipids: a systematic review and meta-analysis of randomized controlled trials. J Am Heart Assoc 2015;4(10):e002408. https://doi.org/10.1161/JAHA.115.002408.

[14] Sears B, Lawren B. Enter the zone. Manhattan, United States: HarperCollins; 1995.

[15] Khan SS, Ning H, Wilkins JT, Allen N, Carnethon M, Berry JD, et al. Association of body mass index with lifetime risk of cardiovascular disease and compression of morbidity. JAMA Cardiol 2018;3(4):280. https://doi.org/10.1001/jamacardio.2018.0022.

[16] Fontana L, Hu FB. Optimal body weight for health and longevity: bridging basic, clinical, and population research. Aging Cell 2014;13(3):391–400. https://doi.org/10.1111/acel.12207.

[17] National Institutes of Health. Weight management techniques. In: The practical guide: identification, evaluation, and treatment of overweight and obesity in adults; 2000, p. 26–7.

[18] Bangalore S, Fayyad R, Laskey R, DA DM, Messerli FH, Waters DD. Body-weight fluctuations and outcomes in coronary disease. N Engl J Med 2017;376(14):1332–40.

[19] Jensen MD, Ryan DH, Apovian CM, Ard JD, Comuzzie AG, Donato KA, et al. 2013 AHA/ACC/TOS guideline for the management of overweight and obesity in adults: a report of the American College of Cardiology/American Heart Association Task Force on Practice Guidelines and The Obesity Society. Circulation 2014;129(25 Suppl. 2):S102–38. https://doi.org/10.1161/01.cir.0000437739.71477.ee.

[20] WHO Expert Consultation. Appropriate Body-Mass Index for Asian populations and its implications for policy and intervention strategies. Lancet 2004;363(9403):157–63. https://doi.org/10.1016/S0140-6736(03)15268-3.

[21] Rush EC, Goedecke JH, Jennings C, Micklesfield L, Dugas L, Lambert EV, Plank LD. BMI, fat and muscle differences in urban women of five ethnicities from two countries. Int J Obes 2007;31(8):1232–9. https://doi.org/10.1038/sj.ijo.0803576.

[22] Aloia JF, Vaswani A, Mikhail M, Flaster ER. Body composition by dual-energy X-ray absorptiometry in black compared with white women. Osteoporos Int 1999;10(2):114–9. https://doi.org/10.1007/s001980050204.

[23] Chiu M, Austin PC, Manuel DG, Shah BR, Tu JV. Deriving ethnic-specific BMI cutoff points for assessing diabetes risk. Diabetes Care 2011;34(8):1741–8. https://doi.org/10.2337/dc10-2300.

[24]   Johnston BC, Kanters S, Bandayrel K, Wu P, Naji F, Siemieniuk RA, et al. Comparison of weight loss among named diet programs in overweight and obese adults. JAMA 2014;312(9):923. https://doi.org/10.1001/jama.2014.10397.

[25]   Mozaffarian D, Hao T, Rimm EB, Willett WC, Hu FB. Changes in diet and lifestyle and long-term weight gain in women and men. N Engl J Med 2011;364(25):2392–404. https://doi.org/10.1056/NEJMoa1014296.

[26]   Anton SD, Hida A, Heekin K, Sowalsky K, Karabetian C, Mutchie H, et al. Effects of popular diets without specific calorie targets on weight loss outcomes: systematic review of findings from clinical trials. Nutrients 2017;9(8):822. https://doi.org/10.3390/nu9080822.

[27]   Mansoor N, Vinknes KJ, Veierød MB, Retterstøl K. Effects of low-carbohydrate diets v. low-fat diets on body weight and cardiovascular risk factors: a meta-analysis of randomised controlled trials. Br J Nutr 2016;115(3):466–79. https://doi.org/10.1017/S0007114515004699.

[28]   Jenkins DJA, Wong JMW, Kendall CWC, Esfahani A, Ng VWY, Leong TCK, et al. Effect of a 6-month vegan low-carbohydrate ("Eco-Atkins") diet on cardiovascular risk factors and body weight in hyperlipidaemic adults: a randomised controlled trial. BMJ Open 2014;4(2):e003505. https://doi.org/10.1136/bmjopen-2013-003505.

[29]   Jenkins DJA, Jones PJH, Lamarche B, Kendall CWC, Faulkner D, Cermakova L, et al. Effect of a dietary portfolio of cholesterol-lowering foods given at 2 levels of intensity of dietary advice on serum lipids in hyperlipidemia: a randomized controlled trial. JAMA 2011;306(8):831–9. https://doi.org/10.1001/jama.2011.1202.

[30]   Turner-McGrievy GM, Jenkins DJA, Barnard ND, Cohen J, Gloede L, Green AA. Decreases in dietary glycemic index are related to weight loss among individuals following therapeutic diets for type 2 diabetes. J Nutr 2011;141(8):1469–74. https://doi.org/10.3945/jn.111.140921.

[31]   Jenkins DJA, Kendall CWC, McKeown-Eyssen G, Josse RG, Silverberg J, Booth GL, et al. Effect of a low–glycemic index or a high–cereal fiber diet on type 2 diabetes. JAMA 2008;300(23):2742. https://doi.org/10.1001/jama.2008.808.

[32]   Willett W, Manson J, Liu S. Glycemic index, glycemic load, and risk of type 2 diabetes. Am J Clin Nutr 2002;76(1):274S–80S. https://doi.org/10.1093/ajcn/76.1.274S.

[33]   Bhupathiraju SN, Tobias DK, Malik VS, Pan A, Hruby A, Manson JE, et al. Glycemic index, glycemic load, and risk of type 2 diabetes: results from 3 large US cohorts and an updated meta-analysis. Am J Clin Nutr 2014;100(1):218–32. https://doi.org/10.3945/ajcn.113.079533.

[34]   The University of Sydney. Glycemic Index; 2017. Available from: http://www.glycemicindex.com/ [Accessed 11 August 2018].

[35]   Elhayany A, Lustman A, Abel R, Attal-Singer J, Vinker S. A low carbohydrate Mediterranean diet improves cardiovascular risk factors and diabetes control among overweight patients with type 2 diabetes mellitus: a 1-year prospective randomized intervention study. Diabetes Obes Metab 2010;12(3):204–9. https://doi.org/10.1111/j.1463-1326.2009.01151.x.

[36]   Austel A, Ranke C, Wagner N, Görge J, Ellrott T. Weight loss with a modified Mediterranean-type diet using fat modification: a randomized controlled trial. Eur J Clin Nutr 2015;69(8):878–84. https://doi.org/10.1038/ejcn.2015.11.

[37]   Gardner CD, Kiazand A, Alhassan S, Kim S, Stafford RS, Balise RR, et al. Comparison of the Atkins, Zone, Ornish, and LEARN diets for change in weight and related risk factors among overweight premenopausal women. JAMA 2007;297(9):969. https://doi.org/10.1001/jama.297.9.969.

[38]   Poulsen SK, Due A, Jordy AB, Kiens B, Stark KD, Stender S, et al. Health effect of the New Nordic Diet in adults with increased waist circumference: a 6-mo randomized controlled trial. Am J Clin Nutr 2014;99(1):35–45. https://doi.org/10.3945/ajcn.113.069393.

[39]   Huang R-Y, Huang C-C, Hu FB, Chavarro JE. Vegetarian diets and weight reduction: a meta-analysis of randomized controlled trials. J Gen Intern Med 2016;31(1):109–16. https://doi.org/10.1007/s11606-015-3390-7.

[40]   Rapsomaniki E, Timmis A, George J, Pujades-Rodriguez M, Shah AD, Denaxas S, et al. Blood pressure and incidence of twelve cardiovascular diseases: lifetime risks, healthy life-years lost, and age-specific associations in 1·25 million people. Lancet 2014;383(9932):1899–911. https://doi.org/10.1016/S0140-6736(14)60685-1.

[41]   Schnall PL, Schwartz JE, Landsbergis PA, Warren K, Pickering TG, Devereux RB. Relation between job strain, alcohol, and ambulatory blood pressure. Hypertension 1992;19(5):488–94. https://doi.org/10.1161/01.hyp.26.3.413.

[42]   Casey DE, Collins KJ, Dennison Himmelfarb C, DePalma SM, Faha M, Gidding S, Jamerson KA, et al. 2017 ACC/AHA/AAPA/ABC/ACPM/AGS/APhA/ASH/ASPC/NMA/PCNA Guideline for the prevention, detection, evaluation, and management of high blood pressure in adults: a report of the American College of Cardiology/American Heart Association Task Force on Clinical Practice Guidelines. J Am Coll Cardiol 2018;71:e127–248. https://doi.org/10.1016/j.jacc.2017.11.006.

[43] Sheps SG. Pulse pressure: an indicator of heart health? Mayo Clinic; 2016. Available from: https://www.mayoclinic.org/diseases-conditions/high-blood-pressure/expert-answers/pulse-pressure/faq-20058189. [Accessed 31 December 2018].

[44] Whelton PK, Carey RM, Aronow WS, Casey DE, Collins KJ, Dennison Himmelfarb C, et al. 2017 ACC/AHA/AAPA/ABC/ACPM/AGS/APhA/ASH/ASPC/NMA/PCNA guideline for the prevention, detection, evaluation, and management of high blood pressure in adults: executive summary. Hypertension 2017; https://doi.org/10.1161/HYP.0000000000000066.

[45] Selassie A, Wagner CS, Laken ML, Ferguson ML, Ferdinand KC, Egan BM. Progression is accelerated from prehypertension to hypertension in blacks. Hypertension 2011;58(4):579–87. https://doi.org/10.1161/HYPERTENSIONAHA.111.177410.

[46] Franklin SS, Larson MG, Khan SA, Wong ND, Leip EP, Kannel WB, Levy D. Does the relation of blood pressure to coronary heart disease risk change with aging? The Framingham Heart Study. Circulation 2001;103(9):1245–9.

[47] Wang N-Y, Young JH, Meoni LA, Ford DE, Erlinger TP, Klag MJ. Blood pressure change and risk of hypertension associated with parental hypertension: The Johns Hopkins Precursors Study. Arch Intern Med 2008;168(6):643. https://doi.org/10.1001/archinte.168.6.643.

[48] Root MM, Dawson HR. DASH-like diets high in protein or monounsaturated fats improve metabolic syndrome and calculated vascular risk. Int J Vitam Nutr Res 2013;83(4):224–31. https://doi.org/10.1024/0300-9831/a000164.

[49] Svetkey LP, Sacks FM, Obarzanek E, Vollmer WM, Appel LJ, Lin PH, et al. The DASH diet, sodium intake and blood pressure trial (DASH-sodium): rationale and design. DASH-Sodium Collaborative Research Group. J Am Diet Assoc 1999;99(8 Suppl):S96–104.

[50] Institute of Medicine. (2013). Sodium Intake in Populations.

[51] National Academies of Sciences Engineering and Medicine. Consensus study report: dietary reference intakes for sodium and potassium; 2019. Available from: https://www.nap.edu/resource/25353/030519DRISodiumPotassium.pdf.

[52] Song EK, Moser DK, Dunbar SB, Pressler SJ, Lennie TA. Dietary sodium restriction below 2 g per day predicted shorter event-free survival in patients with mild heart failure. Eur J Cardiovasc Nurs 2014;13(6):541–8. https://doi.org/10.1177/1474515113517574.

[53] Taylor RS, Ashton KE, Moxham T, Hooper L, Ebrahim S. Reduced dietary salt for the prevention of cardiovascular disease: a meta-analysis of randomized controlled trials (Cochrane review). Am J Hypertens 2011;24(8):843–53. https://doi.org/10.1038/ajh.2011.115.

[54] Jenkins DJA, Jones PJ, Frohlich J, Lamarche B, Ireland C, Nishi SK, et al. The effect of a dietary portfolio compared to a DASH-type diet on blood pressure. Nutr Metab Cardiovasc Dis 2015;25(12):1132–9. https://doi.org/10.1016/j.numecd.2015.08.006.

[55] Hernáez Á, Castañer O, Elosua R, Pintó X, Estruch R, Salas-Salvadó J, et al. Mediterranean diet improves high-density lipoprotein function in high-cardiovascular-risk individualsclinical perspective. Circulation 2017;135(7):633–43. https://doi.org/10.1161/CIRCULATIONAHA.116.023712.

[56] Evans CE, Greenwood DC, Threapleton DE, Gale CP, Cleghorn CL, Burley VJ. Glycemic index, glycemic load, and blood pressure: a systematic review and meta-analysis of randomized controlled trials. Am J Clin Nutr 2017;105(5):1176–90. https://doi.org/10.3945/ajcn.116.143685.

[57] Ndanuko RN, Tapsell LC, Charlton KE, Neale EP, Batterham MJ. Dietary patterns and blood pressure in adults: a systematic review and meta-analysis of randomized controlled trials. Adv Nutr 2016;7(1):76–89. https://doi.org/10.3945/an.115.009753.

[58] Wightman JD, Heuberger RA. Effect of grape and other berries on cardiovascular health. J Sci Food Agric 2015;95(8):1584–97. https://doi.org/10.1002/jsfa.6890.

[59] Reynolds K, Chin A, Lees KA, Nguyen A, Bujnowski D, He J. A meta-analysis of the effect of soy protein supplementation on serum lipids. Am J Cardiol 2006;98(5):633–40. https://doi.org/10.1016/j.amjcard.2006.03.042.

[60] Khan K, Jovanovski E, Ho HVT, Marques ACR, Zurbau A, Mejia SB, et al. The effect of viscous soluble fiber on blood pressure: a systematic review and meta-analysis of randomized controlled trials. Nutr Metab Cardiovasc Dis 2018;28(1):3–13. https://doi.org/10.1016/j.numecd.2017.09.007.

[61] Downs JR, Clearfield M, Weis S, Whitney E, Shapiro DR, Beere PA, et al. Primary prevention of acute coronary events with lovastatin in men and women with average cholesterol levels: results of AFCAPS/TexCAPS. Air Force/Texas Coronary Atherosclerosis Prevention Study. JAMA 1998;279(20):1615–22.

[62] Ford I, Murray H, McCowan C, Packard CJ. Long-term safety and efficacy of lowering low-density lipoprotein cholesterol with statin therapy 20-year follow-up of west of Scotland coronary prevention study. Circulation 2016;133(11):1073–80. https://doi.org/10.1161/CIRCULATIONAHA.115.019014.

[63] Ford I, Murray H, Packard CJ, Shepherd J, Macfarlane PW, Cobbe SM. West of Scotland Coronary Prevention Study Group. Long-term follow-up of the West of Scotland Coronary Prevention Study. N Engl J Med 2007;357(15):1477–86. https://doi.org/10.1056/NEJMoa065994.

[64] Montori VM, Devereaux PJ, Adhikari NKJ, Burns KEA, Eggert CH, Briel M, et al. Randomized trials stopped early for benefit: a systematic review. JAMA 2005;294(17):2203–9. https://doi.org/10.1001/jama.294.17.2203.

[65] Packard CJ, Ford I. Long-term follow-up of lipid-lowering trials. Curr Opin Lipidol 2015;26(6):572–9. https://doi.org/10.1097/MOL.0000000000000230.

[66] Ridker PM, Danielson E, Fonseca FAH, Genest J, Gotto AM, Kastelein JJP, et al. Rosuvastatin to prevent vascular events in men and women with elevated C-reactive protein. N Engl J Med 2008;359(21):2195–207. https://doi.org/10.1056/NEJMoa0807646.

[67] Sever PS, Dahlöf B, Poulter NR, Wedel H, Beevers G, Caulfield M, et al. Prevention of coronary and stroke events with atorvastatin in hypertensive patients who have average or lower-than-average cholesterol concentrations, in the Anglo-Scandinavian Cardiac Outcomes Trial–Lipid Lowering Arm (ASCOT-LLA): a multicentre randomis. Drugs 2004;64(Suppl. 2):43–60.

[68] Vallejo-Vaz AJ, Robertson M, Catapano AL, Watts GF, Kastelein JJ, Packard CJ, et al. Low-density lipoprotein cholesterol lowering for the primary prevention of cardiovascular disease among men with primary elevations of low-density lipoprotein cholesterol levels of 190 mg/dL or above: analyses from the WOSCOPS (West of Scotland Coronary P). Circulation 2017;136(20):1878–91. https://doi.org/10.1161/CIRCULATIONAHA.117.027966.

[69] Rosenson RS, Brewer HB, Davidson WS, Fayad ZA, Fuster V, Goldstein J, et al. Cholesterol efflux and atheroprotection: advancing the concept of reverse cholesterol transport. Circulation 2012;125(15):1905–19. https://doi.org/10.1161/CIRCULATIONAHA.111.066589.

[70] Katcher HI, Hill AM, Lanford JLG, Yoo JS, Kris-Etherton PM. Lifestyle approaches and dietary strategies to lower LDL-cholesterol and triglycerides and raise HDL-cholesterol. Endocrinol Metab Clin N Am 2009;38(1):45–78. https://doi.org/10.1016/j.ecl.2008.11.010.

[71] Roussell MA, Kris-Etherton P. Effects of lifestyle interventions on high-density lipoprotein cholesterol levels. J Clin Lipidol 2007;1(1):65–73. https://doi.org/10.1016/j.jacl.2007.02.005.

[72] Walldius G, Jungner I. Apolipoprotein B and apolipoprotein A-I: risk indicators of coronary heart disease and targets for lipid-modifying therapy. J Intern Med 2004;255(2):188–205. Retrieved from, http://www.ncbi.nlm.nih.gov/pubmed/14746556.

[73] Anderson TJ, Grégoire J, Pearson Pharmd GJ, Pharmd AB, Couture P, Dawes M, et al. About this Pocket Guide CCS Dyslipidemia Guidelines Primary Panel; n.d. Available from www.ccs.ca.

[74] Meadows TA, Bhatt DL, Hirsch AT, Creager MA, Califf RM, Ohman EM, et al. Ethnic differences in the prevalence and treatment of cardiovascular risk factors in US outpatients with peripheral arterial disease: insights from the reduction of atherothrombosis for continued health (REACH) registry. Am Heart J 2009;158(6):1038–45. https://doi.org/10.1016/j.ahj.2009.09.014.

[75] Jenkins DJA, Kendall CWC, Marchie A, Faulkner DA, Wong JMW, de Souza R, et al. Direct comparison of a dietary portfolio of cholesterol-lowering foods with a statin in hypercholesterolemic participants. Am J Clin Nutr 2005;81(2):380–7.

[76] Labonté M-È, Jenkins DJA, Lewis GF, Chiavaroli L, Wong JMW, Kendall CWC, et al. Adding MUFA to a dietary portfolio of cholesterol-lowering foods reduces apoAI fractional catabolic rate in subjects with dyslipidaemia. Br J Nutr 2013;110(3):426–36. https://doi.org/10.1017/S000711451200534X.

[77] Adamsson V, Reumark A, Fredriksson I-B, Hammarström E, Vessby B, Johansson G, Risérus U. Effects of a healthy Nordic diet on cardiovascular risk factors in hypercholesterolaemic subjects: a randomized controlled trial (NORDIET). J Intern Med 2011;269(2):150–9. https://doi.org/10.1111/j.1365-2796.2010.02290.x.

[78] Obarzanek E, Sacks FM, Vollmer WM, Bray GA, Miller ER, Lin P-H, et al. Effects on blood lipids of a blood pressure–lowering diet: the Dietary Approaches to Stop Hypertension (DASH) Trial. Am J Clin Nutr 2001;74(1):80–9. https://doi.org/10.1093/ajcn/74.1.80.

[79] Jenkins DJA, Kendall CWC, Faulkner DA, Kemp T, Marchie A, Nguyen TH, et al. Long-term effects of a plant-based dietary portfolio of cholesterol-lowering foods on blood pressure. Eur J Clin Nutr 2008;62(6):781–8. https://doi.org/10.1038/sj.ejcn.1602768.

[80] Dansinger ML, Gleason JA, Griffith JL, Selker HP, Schaefer EJ. Comparison of the Atkins, Ornish, Weight Watchers, and Zone diets for weight loss and heart disease risk reduction: a randomized trial. JAMA 2005;293(1):43–53. https://doi.org/10.1001/jama.293.1.43.

[81] Sullivan S, Samuel S. Effect of short-term pritikin diet therapy on the metabolic syndrome. J Cardiometab Syndr 2006;1(5):308–12. https://doi.org/10.1111/j.1559-4564.2006.05732.x.

[82] Anderson TJ, Grégoire J, Pearson GJ, Barry AR, Couture P, Dawes M, et al. 2016 Canadian Cardiovascular Society guidelines for the management of dyslipidemia for the prevention of cardiovascular disease in the adult. Can J Cardiol 2016;32(11):1263–82. https://doi.org/10.1016/j.cjca.2016.07.510.

[83] Kannel WB, McGee DL. Diabetes and cardiovascular disease. The Framingham study. JAMA 1979;241(19):2035–8.

[84] Ridker PM, Glynn RJ, Hennekens CH. C-reactive protein adds to the predictive value of total and HDL cholesterol in determining risk of first myocardial infarction. Circulation 1998;97(20):2007–11.

[85] Ridker PM, Buring JE, Shih J, Matias M, Hennekens CH. Prospective study of C-reactive protein and the risk of future cardiovascular events among apparently healthy women. Circulation 1998;98(8):731–3.

[86] Jensen MD, Ryan DH, Apovian CM, Ard JD, Comuzzie AG, Donato KA, et al. 2013 AHA/ACC/TOS guideline for the management of overweight and obesity in adults: a report of the American College of Cardiology/American Heart Association Task Force on Practice Guidelines and The Obesity Society. Circulation 2014;129(25 Suppl 2):S102–38.

[87] Jaacks LM, Kapoor D, Singh K, Narayan KMV, Ali MK, Kadir MM, et al. Vegetarianism and cardiometabolic disease risk factors: differences between South Asian and US adults. Nutrition 2016;32(9):975–84. https://doi.org/10.1016/j.nut.2016.02.011.

[88] Chen ST, Maruthur NM, Appel LJ. The effect of dietary patterns on estimated coronary heart disease risk: results from the dietary approaches to stop hypertension (DASH) trial. Circ Cardiovasc Qual Outcomes 2010;3(5):484–9. https://doi.org/10.1161/CIRCOUTCOMES.109.930685.

Chapter 2

Development and Evidence Base

INTRODUCTION

The first Dietary Portfolio pilot study was carried out in 2000. In this study, the cholesterol-lowering effects of Lovastatin, a cholesterol-lowering medication was compared with the effect of a diet consisting of Portfolio foods. At the time of this study, diet-related strategies to lower cholesterol were seen as relatively ineffective [1]. Researchers had only observed a 4%–13% reduction in cholesterol from dietary intervention studies [2–5], compared to 28%–35% cholesterol reduction seen in long-term trials with statins drugs like Lovastatin [6,7].

Despite the effectiveness of statins, some individuals cannot take them certainly at the required dose due to issues with side effects like statin-associated muscle symptoms (SAMS) which results in muscle stiffness, aching, pain and cramps.

Scientific literature at the time indicated that some foods could modestly reduce cholesterol levels. These foods (plant sterols, viscous fiber, soy, and nuts) had also been recognized by health authorities for their ability to reduce cholesterol (shown in Box 2.1). However, when compared to the effects of statin drugs, the cholesterol lowering ability of the foods on their own to reduce cholesterol levels was minimal. No studies existed examining their combined effect.

Individuals may wish to use diet in addition to drug therapy to reduce levels of medication, decrease interactions from multiple medications, and/or increase the overall effectiveness of drug therapy.

This gap in the research led to the idea for the Dietary Portfolio, a collection of foods all of which uniquely work to reduce cholesterol and prevent coronary heart disease (CHD) (see Table 2.1). The name "Dietary Portfolio" was chosen as the diet is supposed to function in a similar way to a stock portfolio where new assets (or foods) are continually collected and added to the diet as new evidence emerges. As a result of this approach, the number of foods included in the Dietary Portfolio will continue to grow as evidence emerges, making it an ever evolving diet, rooted strongly in evidence.

This chapter will detail the key studies that constitute the evidence base of the Dietary Portfolio, as well as later additions and modifications to the diet.

BOX 2.1 First Adopters to Endorse Portfolio Component Foods

- The National Cholesterol Education Program (Adult Treatment Panel III) advocated for the addition of plant sterols (2 g/d) and viscous fiber (10–25 g/d) to the diet [8].
- The American Heart Association acknowledges the benefits of soy protein and nuts as healthy foods [9,10].
- Food and Drug Administration allows health claims related to the reduction of CVD risk through the cholesterol lowering ability of plant sterols, β-glucan fiber, and nuts [11].

The Portfolio Diet for Cardiovascular Disease Risk Reduction. https://doi.org/10.1016/B978-0-12-810510-8.00002-9

TABLE 2.1 Summary of Portfolio Diet Foods and Quantities per 2000 kcal/day (Average Caloric Consumption of an Individual)

Portfolio food	Quantity per day	Example
Nuts and Seeds	45 g	Almonds, walnuts, or peanuts
Plant based proteins	50 g	Edamame, tofu, tempeh, soymilk, textured, soy protein, lentils, beans
Viscous fiber	20 g	Oat bran, barley, psyllium, eggplant, okra
Plant sterols	2 g	Fortified foods such as enriched margarine fruit juices and plant milks[a]

[a]Plant sterols occur naturally in foods such as soybeans however to obtain the recommended amounts enriched foods are required.

KEY POINTS

Study 1. Testing the Concept: Portfolio Foods vs. Lovastatin

- First study to test the combined ability of specific foods (nuts, plant sterols, viscous fibers, soy protein) to lower levels of serum cholesterol. The study specifically looked at low density lipoprotein cholesterol (LDL-C), as a high LDL-C level is considered a major risk factor for developing cardiovascular disease (CVD).
- Demonstrated that consumption of Portfolio diet foods results in reductions in LDL-C to the same degree as Lovastatin.

Study 2. Raising High Density Lipoprotein (The Good) Cholesterol

- Addition of high levels of monounsaturated fatty acids (MUFA) found in nuts, further increases the effectiveness of the Dietary Portfolio by increasing high density lipoprotein cholesterol (HDL-C), thereby improving the ratio of "good" cholesterol to "bad." Low ratios of LDL-C: HDL-C have been shown to reduce risk of CVD.

Study 3. Testing its Real-World Applicability

- Even when study foods are self-selected and when patients are only given minimal instructions, the Dietary Portfolio can achieve meaningful reductions in LDL-C. These reductions are related to the adherence rate for the diet.

Overall, reductions in CVD risk factors seen in these core studies would be expected to be the same or greater in the general population, as the patients who participated in these studies were either already following a heart healthy diet before starting the Portfolio diet trials or being compared to individuals assigned to a heart healthy diet.

Further Updates and Applications of the Dietary Portfolio

- **Decreased Oxidized LDL-C.** Addition of strawberries to the Dietary Portfolio resulted in decreased oxidized LDL-C levels, a type of LDL-C which is more reactive and thought to lead to more arterial damage. High levels of oxidized LDL-C potentially further increase risk of CVD.

- **Plant-Based Portfolio.** The Dietary Portfolio includes only plant-based foods for health, environmental, and ethical reasons.
- **Improved Gylcemic Control.** Although CVD has been the main target of the Dietary Portfolio studies, through enhanced glycemic control, the Portfolio Diet may also have application for those at risk of or who already have type 2 diabetes.
- **Food Component Updates.** "Soy protein" was changed to "plant based protein" to include foods such as legumes. "Seeds" were added to the "Nuts" category to make it more inclusive to those with nut allergies who may not be able to eat tree nuts or peanuts. Berries have been added to the viscous fiber section, the plant sterol pillar has been broadened to include supplements and other plant sterol fortified foods such as orange juice and spreads.

MAJOR DIETARY PORTFOLIO STUDIES

Study 1: Testing the Concept: Portfolio Foods vs. Lovastatin [12]

Effects of a dietary portfolio of cholesterol-lowering foods vs lovastatin on serum lipids and C-reactive protein. Jenkins DJ, Kendall CW, Marchie A, Faulkner DA, Wong JM, de Souza R, Emam A, Parker TL, Vidgen E, Lapsley KG, Trautwein EA, Josse RG, Leiter LA, Connelly PW. JAMA 2003 Jul 23;290(4):502–10.

Context

The original Dietary Portfolio study combined four components (plant sterols, soy protein, viscous fiber, and nuts) which had previously been found to individually decrease cholesterol and risk of CVD. This study was the first to look at the collective effect of these components on cholesterol levels. At the time this study was conducted, statin drugs like Lovastatin were the gold standard in the treatment of high cholesterol.

This study was the first to use a combination of specific foods to reduce cholesterol.

Purpose

To compare the cholesterol-lowering potential of the Dietary Portfolio with that of a statin.

Study Design

The study consisted of a randomized controlled trial, conducted over the course of one month. The trial included 46 otherwise healthy adults with high cholesterol levels.

Intervention

The participants were divided into three treatment groups. These groups consisted of the following:

1. **Control:** A diet low in saturated fat, based on the following:
   - milled whole-wheat cereals
   - low-fat dairy products
   - commonly consumed fruits and vegetables
2. **Statin:** The same diet as 1, + 20 mg/day of Lovastatin.

3. **Dietary Portfolio:** A diet containing Portfolio diet foods (per 1000 kcal/day):
   - plant sterols 1.0 g
   - soy protein 21.4 g
   - viscous fiber 9.8 g
   - almonds 14 g

Amounts were adjusted for caloric intake of the individual to ensure there were no significant changes in participants' body weight and ensure that the effects on cholesterol could be observed independent of weight-loss. All diets in this study did not include meat products and were vegetarian. While the Control and Statin treatment groups consumed low-fat dairy products and eggs, the Portfolio treatment group consumed only eggs, replacing low-fat dairy with soy products. While it is now recommended to consume the Portfolio diet in the context of a plant-based diet (one that does not include dairy or eggs), eggs were included to balance dietary cholesterol intake between the Portfolio treatment group and the Control and Statin groups. Diets containing dairy products and eggs contain cholesterol while a completely plant-based diet does not. There is some evidence to suggest that dietary intake of cholesterol can result in increased cholesterol levels for some individuals. To ensure reductions in cholesterol were due to treatment effects and not differences in dietary cholesterol intake, all treatment groups consumed an equivalent amount of dietary cholesterol and therefore eggs were added to the Portfolio treatment group. While eggs were important to study design, they are not included while following the Portfolio diet for therapy.

Results

Blood Lipids and C-Reactive Protein

LDL-C is known as the "bad" cholesterol due to its strong association with CHD. The Control, Statin and Portfolio groups saw decreases in LDL-C of 8%, 31%, and 29%, respectively. Both the Portfolio group and the Statin group showed statistically significant reductions in LDL-C compared to the control. Reductions in the Statin group and the Portfolio group were not significantly different from each other. This result demonstrated that the Portfolio diet was able to reduce LDL-C levels to the same extent as a stain drug.

For more information on any of the risk factors discussed refer to Chapter 1.

Another factor that is also examined when looking at CHD risk is the ratio of apolipoprotein B:apolipoprotein A1 (Apo B: Apo A1). Apo B is present in LDL-C, while Apo A1 is present in HDL-C, sometimes referred to as "good cholesterol." Just like the LDL-C:HDL-C ratio, a reduction in the ratio of Apo B : Apo A1 indicates an improvement in the lipid profile. Apo B: Apo A1 ratios were found to decrease for those in the Portfolio and Statin treatment groups compared to the Control.

Levels of C-reactive protein, a marker of inflammation, were compared between treatment groups. In both the Statin and the Portfolio diet groups, C-reactive protein was significantly reduced to the same amount. Reductions in C-reactive protein were 10% in the Control, 33% in the Statin, and 28% in the Portfolio group. The Portfolio and Statin groups significantly reduced their levels of C-reactive protein compared to the Control group.

Calculated CHD Risk

CHD risk was calculated using the Framingham risk equation. The Framingham equation uses factors including body weight, blood pressure, and blood lipid levels to calculate an individual's risk of developing CHD in the future. Both the Statin group and the Portfolio group had better predicted outcomes than the

Control by 26% and 25%, respectively. The blood lipid reductions seen in the Statin and Portfolio group were primarily responsible for the decrease in the calculated CHD risk.

Conclusions

This study demonstrated that combining cholesterol-lowering foods into a Dietary Portfolio creates an effective approach for the treatment of high cholesterol. The resulting reduction in CHD risk was found to be comparable to that of a first-generation statin, a result previously unseen using diet alone. It is also important to note that both the Control and Statin treatment groups, to which the Portfolio group was compared, were put on a relatively healthy diet low in saturated fat. If the Portfolio diet was compared to the diet of the general population, the effects might have been even more significant.

Study 2. Raising High-Density Lipoprotein Levels [13]

Adding monounsaturated fatty acids to a dietary portfolio of cholesterol-lowering foods in hypercholesterolemia. Jenkins DJA, Chiavaroli L, Wong JMW, Kendall C, Lewis GF, Vidgen E, Lamarche B. Can Med Assoc J 2010;182(18).

Context

Up until 2010 the Portfolio diet had demonstrated meaningful reductions in Apo B and LDL-C but absolute levels of Apo A1 and HDL-C had been relatively unaffected. The reductions in the LDL-C: HDL-C ratios observed in the studies were mostly driven by the reductions in LDL-C. High levels of Apo A1 and low ratios of LDL-C:HDL-C are associated with reduced risk of CHD and therefore important targets for prevention strategies. Evidence from research surrounding the Mediterranean Diet indicated that the reductions seen in the LDL-C: HDL-C ratio and the subsequent increase in Apo A1 can be attributed to the high concentration of monounsaturated fatty acids (MUFA) in the form of olive oil and nuts in the diet [14]. This effect is particularly true when carbohydrate is replaced with MUFA. Evidence found in cohort studies also supported this finding, showing that individuals consuming diets high in MUFA from nuts and vegetable oil had a reduced incidence of CVD [15].

CVD risk factors pt. 1
High levels of LDL-C accompanied by low levels of HDL-C have been shown to be strong indicators of heart disease risk.

CVD risk factors pt. 2
Apo B is a protein found bound to LDL-C, while Apo A1 is a protein found bound to HDL-C. High levels of Apo B and low levels of Apo A1 are associated with cardiac events and are believed to be better indicators than cholesterol alone.

Purpose

To determine if adding MUFA to the Dietary Portfolio would increase its effectiveness in reducing CVD risk factors.

Study Design

This study was a randomized controlled trial conducted over the course of two months with a one-month pre-study period where patients consumed a diet low in saturated fat. This design enabled all participants to be at the same starting point, in terms of diet history, when the trial began. A total of 24 patients with elevated cholesterol levels participated, making it a small but intensive study.

Intervention

Participants were randomly assigned to a Portfolio diet low in MUFA, a Portfolio diet high in MUFA, or a control diet low in saturated fat. In the high MUFA Portfolio arm, 13% of dietary calories in the form of carbohydrate were replaced with MUFA derived from high oleic sunflower oil. The high MUFA Dietary Portfolio (per 1000 kcal/day):

- 1 g plant sterols
- 10.3 g of viscous fiber provided by oats psyllium, eggplant, and okra
- 20 g of soy protein in the form of tofu, soy meat analogs, and soymilk
- 21.5 g of almonds

CVD risk factors pt. 3
LDL-C and HDL-C ratios are still more commonly used indicators of CVD as they are better understood by health care professionals and the public than ratios of Apo B: Apo A1.

Results

Blood Lipids and C-Reactive Protein

HDL-C and Apo A1, a component of HDL-C were significantly higher in the treatment group consuming the high MUFA Portfolio diet compared to the group consuming the original Portfolio diet low in MUFA, with a treatment difference of 13 % for HDL-C and 10% for Apo A1. These results suggest that adding MUFA to the Portfolio diet can raise good cholesterol.

The ratio of total cholesterol to HDL-C is a significant predictor of CVD. Significant reductions were seen in the ratio between total cholesterol and HDL-C with a treatment difference of −7% and in the ratio of Apo A1: Apo B with a treatment difference of −5.1 %.

Similar to the previous study, both Dietary Portfolio groups (low and high MUFA) showed significant reductions in LDL-C of −19% in the Low MUFA portfolio group and −21% in the High MUFA portfolio group compared to the control. Apo B was also reduced in both Portfolio treatment groups: −18% in the low MUFA portfolio group and −16% MUFA in the high MUFA portfolio group.

In addition, C-reactive protein was significantly reduced in the high MUFA Portfolio diet group by 77% compared to the low MUFA Portfolio diet group.

Role of C-reactive protein
C-reactive protein, a marker of inflammation, has been demonstrated to be an important marker of CVD risk. Reductions in C-reactive protein with antiinflammatory drugs reduces CVD and cancer deaths.

Conclusions

The addition of moderate amounts of MUFA to the Portfolio diet significantly increased its potential to reduce the risk of CHD by increasing HDL-C levels. The 13% increase in HDL-C achieved with the Portfolio diet high in MUFA was greater than the increases commonly observed with gemfibrozil, a drug often used to treat hyperlipidemia (elevated blood lipid levels). Gemfibrozil, in some studies, has achieved increases in HDL-C of 6% and has been shown to reduce the relative risk of cardiovascular disease by 22% [17]. This study has shown that, in combination with other Portfolio diet foods, MUFA further

Update on MUFA
Recently the PREDIMED study from Spain with 7000 high risk individuals randomized to either a control or a Mediterranean diet with nuts or olive oil (MUFA rich foods) showed a 30% risk reduction in CVD over ~4 year period compared to the control [16].

increases the effectiveness of the Portfolio diet making it competitive with drug-based treatments for CVD. Again, the Portfolio diet with high MUFA content is expected to achieve better results in the general population as participants in both the control group and the Portfolio diet with low MUFA treatment group were all on a diet low in saturated fat prior to the start of the study.

Study 3. Testing Its Real-World Applicability [18]

Effect of a dietary portfolio of cholesterol-lowering foods given at 2 levels of intensity of dietary advice on serum lipids in hyperlipidemia.
Jenkins DJA, Jones PJH, Lamarche B, Kendall CWC, Faulkner D, Cermakova L, … Frohlich J. JAMA 2011;306(8):831–9.

Context

After the positive results of the past Portfolio diet studies, researchers were interested in more thoroughly testing the Portfolio diet. No long-term studies had been undertaken at this time and the studies that had been done were metabolic studies with all the food provided. It was important to understand if the Portfolio diet was a feasible diet to ask individuals to follow when they had to purchase the foods themselves.

Metabolic diets

One of the benefits of metabolic studies is that they are tightly controlled making people much more likely to adhere to a diet. They demonstrate the maximum potential of dietary interventions. However, they do not reflect real world conditions, making it important to test diets when food is not provided.

Purpose

To investigate the long-term effects of Portfolio diet foods under varying conditions. Specifically, to investigate the effect of offering varying levels of guidance on the success of the diet. This testing was carried out at multiple locations to broaden the applicability of the findings.

Study Design

This study was a randomized controlled study, carried out between 2007 and 2009. It included 351 participants with hyperlipidemia from four treatment centers across Canada (Toronto, Vancouver, Winnipeg, and Quebec City). Treatment was carried out over the course of 6 months.

Intervention

Three different dietary advice paradigms were used:

1. **Control:** advice on maintaining a diet low in saturated fat.
2. **Routine advice:** routine advice on adhering to a Portfolio diet, administered with only two sessions in the 6-month period.
3. **Intensive advice:** intensive counseling with advice on adhering to a Portfolio diet administered in seven counseling sessions in the 6-month period.

The control treatment group diet consisted of

- low-fat dairy
- whole grain cereals
- fruits and vegetables with specific avoidance of Portfolio diet components

The two Portfolio diet treatment groups were recommended to consume a similar diet as in the metabolically controlled studies (per 1000 kcal/day):

- 0.94 g of plant sterols in the form of plant sterol ester-enriched margarine
- 9.8 g of viscous fiber from oats, barley, and psyllium
- 22.5 g of soy protein from soy milk, tofu, and soy meat analogues
- 22.5 g of nuts
- while no amount was specified, participants were advised to consume peas and beans such as lentils and chickpeas

Results

Adherence

Adherence to the Dietary Portfolio was somewhat better among the intensive Portfolio diet group at 46% but was not significantly different from the routine Portfolio Diet treatment group which had a 41% adherence. As expected, the more an individual adhered to the diet the greater their reductions in LDL-C. As one would expect, the adherence was lower when compared to the previous metabolically controlled studies, however not outside the norm for dietary intervention studies of this nature.

Blood Lipids

Both Portfolio diet treatment groups significantly reduced their LDL-C levels and both Portfolio diet treatment groups saw significantly greater reductions in LDL-C compared to the control. At the end of the 6 months, both Portfolio groups significantly reduced the LDL-C:HDL-C ratio and both were significantly lower than the control.

CHD Risk

Using the Framingham equation, CHD risk was reduced by 11% for the routine Portfolio diet group. This was similar to the risk reduction seen with the intensive Portfolio diet treatment group of −11%. While not significantly different from each other they were significantly different from the control which saw no significant reduction in relative risk of CHD (−.05%).

Treatment Center Differences

The percent reduction in LDL-C was significantly greater at the Vancouver site compared to the Portfolio diet treatment groups in Toronto or Winnipeg. The greater reductions in LDL-C seen at the Vancouver site were related to the significantly higher adherence rates at that site.

Food environment pt. 1

One of the reasons that Vancouver was different from Toronto and Winnipeg may have been that the number of health food stores and vegetarian restaurants per 100,000 people is much greater in Vancouver than the other two cities. This underscores the importance of the environment on compliance and effectiveness of dietary interventions.

Conclusions

This study demonstrated that the routine Portfolio diet treatment group was able to achieve significant reductions in LDL-C in 6 months, with minimal instruction. Furthermore, the reduction in LDL-C was not significantly different from that achieved using an intensive counseling scenario, indicating that it is realistic to expect people to adopt the Portfolio diet even when only minimal instruction is provided.

Putting It All Together: A Systematic Review and Meta-analysis

In 2018 a systemic review and meta-analysis on Portfolio diet studies was published in Progress in Cardiovascular Diseases [19]. The purpose of a systematic review and meta-analysis is to summarize and analyze all the available evidence of an intervention using specified criteria. They are useful as they can combine all available results and therefore give an estimate of the overall effect of the intervention and how variable it is between studies. This review included all studies which compared the effect of a Portfolio dietary pattern with an energy matched control diet devoid of Portfolio foods on cardiometabolic risk factors. A total of seven trial comparisons in 439 participants were included in the final analysis. The Portfolio dietary pattern significantly reduced LDL-C levels by about 17% compared to the control diet. Non-HDL-C, Apo-B, C-reactive protein and estimated 10-year CVD risk were also reduced on the Portfolio diet. This review therefore supports the use of the Dietary Portfolio as a means to reduce LDL-C levels and risk of developing CVD. See Box 2.2 for a list of organizations that currently recommend the Portfolio Diet.

Food environment pt. 2

Patients accepted into this study were consuming diets low in saturated fat prior to the start of the study and represent a population for whom standard therapeutic dietary advice had not achieved the desired targets. In light of this, the paper hypothesizes that reductions in the general population may be even greater than those seen in this study providing the environment surrounding them is conducive to dietary change. Part of the aim of our book therefore is to create that "conducive environment."

BOX 2.2 International Guidelines Recommending the Dietary Portfolio

- Canadian Cardiovascular Society Guidelines [20,21]
- Diabetes Canada [22]
- European Atherosclerosis Society (EAS) [23]
- Heart UK [24,25]

Further Updates and Applications of the Dietary Portfolio
Strawberries and Oxidized LDL-C Reduction

28 subjects who had been following the Portfolio diet for a mean of 2.5 years were randomized to receive either supplements of strawberries (454 g/day, 112 kcal) or additional oat bran bread (65 g/d, 112 kcal, ≈2 g β-glucan) (control) for one month. After one month participants who were assigned to the strawberry group switched to the control group and vice versa. Addition of strawberries resulted in significant reductions in oxidized LDL-C levels compared to control. This reduction is significant as oxidized LDL-C is more reactive than regular LDL-C and is thought to lead to more arterial damage. The results of this study demonstrated that strawberry supplementation increased the palatability of the diet, reduced the oxidative damage to LDL-C while not changing the overall LDL-C reductions. Supplementing the Portfolio diet with berries such as strawberries, may be a helpful strategy to improve the Portfolio's effectiveness and acceptance.

A Plant-Based Portfolio Diet

Although the original studies examining the Portfolio diet included animal protein coming from egg, the current Dietary Portfolio does not include any animal products. This decision was primarily driven by health, environmental, and ethical concerns which will be further elaborated upon in later sections. Plant-based diets are also practical considering the large quantities of specific foods required to follow the dietary portfolio. After you finish loading up your plate with fruits, vegetables, nuts, and soy there is not much space for anything else. That being said, there is still plenty of room for customization of the existing components. The dietary portfolio can easily be adapted to a wide range of ethnic and culturally specific lifestyles.

The Portfolio diet is recommended to be eaten in the context of a vegetarian or vegan diet or with minimal animal product consumption.

Enhanced Portfolio

As the name 'Portfolio' suggests, the number and types of foods to be included into the Portfolio pattern will expand as new scientific evidence emerges. To include the most recent developments the term "Enhanced Portfolio" is used to indicate the updated version [19]. More information on this is available on the St Michael's Hospital website.[1] In the Enhanced Portfolio, some of the original pillars (nuts, soy, viscous fiber and plant sterols) have been renamed or broadened to allow inclusion of the new foods. The Nut pillar now also includes seeds and the soy pillar is now named "plant protein" to include both soy products and legumes. Berries, like strawberries, have been added to the viscous fiber section, the plant sterol pillar has been broadened to include supplements and other plant sterol fortified foods such as orange juice and spreads.

Applications Beyond CVD

The Portfolio diet may aid in the treatment of type 2 diabetes. Diets high in fiber and nuts, such as the Portfolio diet, have been shown to benefit glycemic control for those with type 2 diabetes by reducing the glycemic index and glycemic load (GI/GL) [26–28]. In long-term studies, nuts, and fiber have shown to reduce the incidence of diabetes [29]. A dual approach in addressing both diabetes and CVD risk factors is particularly useful as people with diabetes are 2–4 times more likely to develop CVD [30]. In addition, cholesterol lowering statins have been shown to increase the risk of diabetes in certain phenotypes [31]. This increased risk makes treatments that help reduce or eliminate the need for statins, such as the Portfolio diet, particularly beneficial.

Glycemic control occurs when blood glucose levels after eating are reduced. This reduction is important for people with type 2 diabetes who experience high glucose rises after meals.

---

[1] http://www.stmichaelshospital.com/media/hospital_news/2018/0706.php

REFERENCES

[1] Ramsay LE, Yeo WW, Jackson PR. Dietary reduction of serum cholesterol concentration: time to think again. BMJ (Clinical Research Ed) 1991;303(6808):953–7. Retrieved from, http://www.ncbi.nlm.nih.gov/pubmed/1954418.

[2] Jenkins DJA, Kendall CWC, Jackson C-JC, Connelly PW, Parker T, Faulkner D, et al. Effects of high- and low-isoflavone soyfoods on blood lipids, oxidized LDL, homocysteine, and blood pressure in hyperlipidemic men and women. Am J Clin Nutr 2002;76(2):365–72. Retrieved from http://www.ncbi.nlm.nih.gov/pubmed/12145008.

[3] Jenkins DJA, Kendall CWC, Marchie A, Parker TL, Connelly PW, Qian W, et al. Dose response of almonds on coronary heart disease risk factors: blood lipids, oxidized low-density lipoproteins, lipoprotein(a), homocysteine, and pulmonary nitric oxide: a randomized, controlled, crossover trial. Circulation 2002;106(11):1327–32. Retrieved from http://www.ncbi.nlm.nih.gov/pubmed/12221048.

[4] Law MR. Plant sterol and stanol margarines and health. West J Med 2000;173(1):43–7. Retrieved from, http://www.ncbi.nlm.nih.gov/pubmed/10903294.

[5] Olson BH, Anderson SM, Becker MP, Anderson JW, Hunninghake DB, Jenkins DJ, et al. Psyllium-enriched cereals lower blood total cholesterol and LDL cholesterol, but not HDL cholesterol, in hypercholesterolemic adults: results of a meta-analysis. J Nutr 1997;127(10):1973–80. Retrieved from, http://www.ncbi.nlm.nih.gov/pubmed/9311953.

[6] Downs JR, Clearfield M, Weis S, Whitney E, Shapiro DR, Beere PA, et al. Primary prevention of acute coronary events with lovastatin in men and women with average cholesterol levels: results of AFCAPS/TexCAPS. Air Force/Texas Coronary Atherosclerosis Prevention Study. JAMA 1998;279(20):1615–22. Retrieved from, http://www.ncbi.nlm.nih.gov/pubmed/9613910.

[7] Heart Protection Study Collaborative Group. MRC/BHF Heart Protection Study of cholesterol lowering with simvastatin in 20536 high-risk individuals: a randomised placebocontrolled trial. Lancet 2002;360(9326):7–22. https://doi.org/10.1016/S0140-6736(02)09327-3.

[8] Nayor M, Vasan RS. Recent update to the us cholesterol treatment guidelines: a comparison with international guidelines. Circulation 2016;133(18):1795–806.

[9] Sacks FM, Lichtenstein A, Van Horn L, Harris W, Kris-Etherton P, Winston M. Soy protein, isoflavones, and cardiovascular health: An American Heart Association Science Advisory for professionals from the Nutrition Committee. Circulation 2006; https://doi.org/10.1161/CIRCULATIONAHA.106.171052.

[10] Erdman JW. "AHA Science Advisory: Soy Protein and Cardiovascular Disease: A Statement for Healthcare Professionals from the Nutrition Committee of the AHA." Circulation 2000;102(20):2555–9. http://www.ncbi.nlm.nih.gov/pubmed/11076833.

[11] FDA, n.d. https://www.fda.gov.

[12] Jenkins DJ, Kendall CW, Marchie A, Faulkner DA, Wong JM, de Souza R, et al. Effects of a dietary portfolio of cholesterol-lowering foods vs lovastatin on serum lipids and C-reactive protein. JAMA 2003;290(4):502–10.

[13] Jenkins DJA, Chiavaroli L, Wong JMW, Kendall C, Lewis GF, Vidgen E, et al. Adding monounsaturated fatty acids to a dietary portfolio of cholesterol-lowering foods in hypercholesterolemia. Can Med Assoc J 2010;182(18).

[14] Grundy SM. Comparison of monounsaturated fatty acids and carbohydrates for lowering plasma cholesterol. N Engl J Med 1986;314(12):745–8. https://doi.org/10.1056/NEJM198603203141204.

[15] Mensink RP, Zock PL, Kester ADM, Katan MB. Effects of dietary fatty acids and carbohydrates on the ratio of serum total to HDL cholesterol and on serum lipids and apolipoproteins: a meta-analysis of 60 controlled trials. Am J Clin Nutr 2003;77(5):1146–55. Retrieved from http://www.ncbi.nlm.nih.gov/pubmed/12716665.

[16] Estruch R, Ros E, Salas-Salvadó J, Covas M-I, Corella D, Arós F, et al. Primary Prevention of Cardiovascular Disease with a Mediterranean Diet Supplemented with Extra-Virgin Olive Oil or Nuts. New England Journal of Medicine 2018;378(25):e34. https://doi.org/10.1056/NEJMoa1800389.

[17] Robins SJ, Collins D, Wittes JT, Papademetriou V, Deedwania PC, Schaefer EJ, et al. Relation of gemfibrozil treatment and lipid levels with major coronary events: VA-HIT: a randomized controlled trial. JAMA 2001;285(12):1585–91. Retrieved from, http://www.ncbi.nlm.nih.gov/pubmed/11268266.

[18] Jenkins DJA, Jones PJH, Lamarche B, Kendall CWC, Faulkner D, Cermakova L, et al. Effect of a dietary portfolio of cholesterol-lowering foods given at 2 levels of intensity of dietary advice on serum lipids in hyperlipidemia. JAMA 2011;306(8):831–9.

[19] Chiavaroli L, Nishi SK, Khan TA, Braunstein CR, Glenn AJ, Mejia SB, et al. Portfolio Dietary Pattern and Cardiovascular Disease: A Systematic Review and Meta-analysis of Controlled Trials. Prog Cardiovasc Dis. 2018;61(1):43–53. https://doi.org/10.1016/j.pcad.2018.05.004. Epub 2018 May 26.

[20] Anderson TJ, Gregoire J, Pearson GJ, et al. Canadian cardiovascular society guidelines for the management of dyslipidemia for the prevention of cardiovascular disease in the adult. Can J Cardiol 2016;32:1263–82.

[21] Jenkins DJA, Kendall CWC, Burris L, Sievenpiper JL, Evans MF, Angl EN. The Portfolio Diet: an evidence-based eating plan for lower cholesterol; 2017. Available from: https://www.ccs.ca/images/Images_2017/Portfolio_Diet_Scroll_eng.pdf.

[22] Sievenpiper JL, Chan CB, Dworatzek PD, Freeze C, Williams SL. Nutrition therapy. Can J Diabetes 2018;42(Suppl. 1):S64–S79.

[23] Stroes ES, Thompson PD, Corsini A, et al. Statin-associated muscle symptoms: impact on statin therapy-European Atherosclerosis Society Consensus Panel Statement on Assessment, Aetiology and Management. Eur Heart J 2015;36:1012–22.

[24] Heart UK: The cholesterol charity. Portfolio diet; 2014. Available from: https://heartuk.org.uk/cholesterol-and-diet/six-super-foods-for-lower-cholesterol/portfolio-diet.

[25] Heart UK: The cholesterol charity. Portfolio diet; 2014. Available from: https://heartuk.org.uk/files/uploads/documents/huk_fs_d01_theportfoliodiet.pdf.

[26] Jenkins AL, Jenkins DJA, Wolever TMS, Rogovik AL, Jovanovski E, Bozikov V, et al. Comparable postprandial glucose reductions with viscous fiber blend enriched biscuits in healthy subjects and patients with diabetes mellitus: acute randomized controlled clinical trial. Croat Med J 2008;49(6):772–82. https://doi.org/10.3325/CMJ.2008.49.722.

[27] Jenkins DJA, Kendall CWC, Marchie A, Faulkner DA, Wong JMW, de Souza R, et al. Direct comparison of a dietary portfolio of cholesterol-lowering foods with a statin in hypercholesterolemic participants. Am J Clin Nutr 2005;81(2):380–7. Retrieved from http://www.ncbi.nlm.nih.gov/pubmed/15699225.

[28] Wolever TM, Jenkins DJ, Vuksan V, Jenkins AL, Buckley GC, Wong GS, Josse RG. Beneficial effect of a low glycaemic index diet in type 2 diabetes. Diabet Med 1992;9(5):451–8. Retrieved from, http://www.ncbi.nlm.nih.gov/pubmed/1611833.

[29] Salas-Salvadó J, Bulló M, Estruch R, Ros E, Covas M-I, Ibarrola-Jurado N, et al. Prevention of diabetes with Mediterranean diets. Ann Intern Med 2014;160(1):1–10. https://doi.org/10.7326/M13-1725.

[30] Kannel WB, McGee DL. Diabetes and cardiovascular disease. The Framingham study. JAMA 1979;241(19):2035–8. Retrieved from http://www.ncbi.nlm.nih.gov/pubmed/430798.

[31] Waters DD, Ho JE, DeMicco DA, Breazna A, Arsenault BJ, Wun C-C, et al. Predictors of new-onset diabetes in patients treated with atorvastatin. J Am Coll Cardiol 2011;57(14):1535–45. https://doi.org/10.1016/j.jacc.2010.10.047.

Chapter 3

How It Works: Mechanisms of Action

INTRODUCTION

Each of the four foods components (plant sterols, viscous fiber, nuts, and soy) included in the original Dietary Portfolio was selected for its unique ability to either lower low density lipoprotein cholesterol (LDL-C) or raise high density lipoprotein cholesterol (HDL-C), and thus reduce the risk of cardiovascular disease (CVD). While the individual effects of each of the four food components have previously been studied, the Dietary Portfolio was the first to combine all these factors in a single diet and investigate their collective effect on cholesterol. As many pathways are utilized by these food components to reduce the risk of CVD, when combined within a single diet their individual ability to reduce CVD risk was proposed to sum, with each food adding its own protective effect. As new evidence emerges, new foods are added to the Portfolio diet. In the latest version of the Portfolio diet, known as the "Enhanced Portfolio" seeds have been added under the heading of nuts to provide an alternative for those with allergies to tree nuts or peanuts. The category of soy has also been expanded to include pulses under the new heading of plant-based protein.

The following chapter explores the evidence base that lead to the inclusion of these foods in the Dietary Portfolio. The primary focus will be on the mechanisms through which each food component works to either lower low density lipoprotein cholesterol (LDL-C) or raise high density lipoprotein cholesterol (HDL-C). This chapter will also briefly outline other health benefits related to these foods, as well as discuss some myths and misconceptions surrounding the foods. A brief summary is provided below. A more detailed explanation of terms and mechanisms can be found within the chapter.

Key Points

Plant sterols

Plant sterols are proposed to reduce cholesterol absorption in the gut, resulting in LDL-C reductions. Many different mechanisms of action have been proposed to account for the ability of plant sterols to reduce LDL-C. The most prominent involves competition for inclusion into the micelles, the process by which cholesterol enters the cells of the intestine and is eventually parceled into very low density lipoprotein (VLDL) and LDL-C by the liver. By displacing cholesterol in the micelles plant sterols prevent LDL-C and VLDL formation.

The Portfolio Diet for Cardiovascular Disease Risk Reduction. https://doi.org/10.1016/B978-0-12-810510-8.00003-0

Fiber

The Dietary Portfolio contains many different types of fiber, however the reduction in LDL-C can be largely attributed to viscous fiber. There are three main mechanisms by which viscous fiber reduces LDL-C.

- Viscous fiber forms a viscous or thick and sticky gel in the intestine. This gel has been found to delay absorption of nutrients, like glucose. A slower release of glucose means that not as much insulin is needed to transport glucose into cells where it can be stored or used for energy. The release of insulin is hypothesized to stimulate the production of cholesterol and therefore reductions in insulin may decrease cholesterol synthesis.
- The thick gel that viscous fiber forms may also trap bile salts and prevent their re-uptake in the terminal ileum. The liver must then use the existing LDL-C to make bile acids and replenish the bile acid pool, thus reducing LDL-C.
- Viscous fiber also interferes with the production of micelles, preventing cholesterol from being absorbed into the cell.
- Viscous fiber increases the production of short chain fatty acids (SCFA) through fermentation in the colon. Specific SCFA have been linked to reduced risk of CVD [1].

Additional benefits of viscous fiber, including psyllium, consumption have been proposed and include; management of glucose levels, prevention of colon cancer irritable bowel syndrome, hemorrhoids, constipation, diarrhea, and applications for weight-loss.

Nuts

Nuts and seeds contain monounsaturated fatty acids (MUFA), plant sterols, vegetable protein, fiber, and phytochemicals all of which may contribute to their cholesterol lowering properties. The addition of nuts such as almonds, hazelnuts, and walnuts to the Dietary Portfolio has been linked with an increase in apolipoprotein A1 (Apo A1) and HDL-C and a decrease in C-reactive protein, and LDL-C, changes which have been shown to decrease the risk of CVD. This chapter primarily focuses on the mechanisms of MUFA and phytochemicals plant sterols, vegetable protein, and fiber in lowering LDL-C and raising HDL-C are discussed in other sections of this chapter. The proposed mechanisms are as follows:

- Replacing carbohydrate intake with MUFA consumption is believed to increase the presence of HDL or "good" cholesterol by making it denser, preventing it from being broken down as easily and allowing it to stay in circulation longer.
- MUFA is also believed to decrease inflammation by reducing the production of inflammatory (C-reactive protein) compounds and increasing the production of antiinflammatory (Apo A1) compounds.

Additional benefits include improved glycemic control for individuals with type 2 diabetes, better fasting glucose levels, lower blood triglycerides, and waist circumference reduction when the equivalent amount of carbohydrate was replaced with nuts in the diet. If tree nuts and peanuts cannot be consumed due to allergies, seeds can be used as a substitute.

Plant-based Protein

Pulses, soy milk and soy meat analogs are high in protein and low in fat. In the context of the Portfolio diet they have been shown to contribute to reductions in LDL-C.

- The bioactive peptides found in soy, may in part be responsible for the observed reductions in LDL-C through interference with LDL-C receptors and improving bile acid regulation.

- Pulses contain viscous fiber which may decrease cholesterol by increasing the excretion of bile acids/salts, slowing glucose absorption, reduce insulin levels and by increasing the production of SCFA.
- Pulses also contain polyphenols, some of which may decrease the risk of CVD by protecting against LDL-C oxidation.
- The role that both soy and pulses play in replacing other food sources that are less heart healthy may be equally important in terms of LDL reduction.

Additional benefits of soy and pulses include: improved glycemic control and weight-loss. Soy on its own may reduce blood pressure.

PLANT STEROLS

Background

Plant sterols are present in all types of vegetable oils, but have been incorporated into the Portfolio diet in the form of sterol-enriched margarine to allow for precise measurement of dosage in a concentrated form and high levels of intake. The most common commercial sources of plant sterols are in enriched margarine with the plant sterols coming from soybean oil or pine tree oil, also known as tall oil. The main plant sterol compound in both soybean and tall oil is β-sitosterol. However, soybean oil has a more diverse sterol profile containing significant amounts of campesterol and stigmasterol. The enriched margarine used in the Portfolio diet studies (Flora Pro-Active Upfield, London, United Kingdom which is now sold under the name Becel in North America) was enriched with plant sterols from soybeans. When 2 g of plant sterols were consumed each day in the context of the Portfolio diet reductions in LDL-C were observed.

> Plant sterols are found in all vegetable oils.

Mechanisms

Plant sterols reduce the absorption of cholesterol in the gut, decreasing serum cholesterol and subsequently, CVD risk. To understand better how plant sterols may block cholesterol absorption it is important to understand the process through which cholesterol is absorbed (Fig. 3.1).

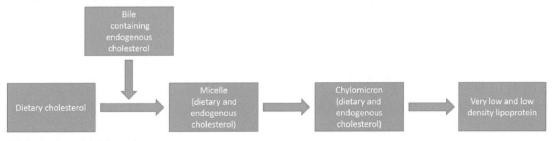

FIG. 3.1   Cholesterol absorption.

Cholesterol Absorption

There are two types of cholesterol: dietary cholesterol and the cholesterol our body makes itself which is called endogenous cholesterol. When we consume foods that contain cholesterol, bile is released to package cholesterol in a format that can be absorbed in the small intestine. This specific format is called a micelle. Interestingly, the bile that the liver produces contains even more cholesterol, but this is endogenous cholesterol, the kind the body makes itself. This endogenous cholesterol is incorporated along with dietary cholesterol into micelles and absorbed by the small intestine for reuse. Once absorbed, both dietary and endogenous cholesterol are packaged into what are known as chylomicrons. Chylomicrons are sent back to the liver after delipidation or triglyceride removal through the circulatory system where they contribute to VLDL and LDL-C. It should be noted that except for a small percentage of the population, most individuals do not absorb a large amount of dietary cholesterol. Most of the cholesterol that is packaged as micelles is endogenous. Therefore, the most effective way to lower cholesterol is to block the reabsorption of endogenous cholesterol, discussed below.

The Role of Plant Sterols in Inhibiting Cholesterol Absorption

Several theories exist to explain how plant sterols inhibit cholesterol re-absorption. The main underpinning of the theories is that it is competition which prevents cholesterol from being absorbed. Plant sterols and cholesterol share a very similar structure making them act in similar ways in the body. In general terms, plant sterols can be thought of as the plant-based version of cholesterol. It is because of their similarity with cholesterol that plant sterols are able to block cholesterol absorption and have a protective effect against CVD. The oldest theory involves competition by plant sterols for incorporation into mixed micelles. Dietary and endogenous cholesterol must first become part of the mixed micelles before they can be absorbed. It was hypothesized that sterols and cholesterol compete for inclusion into the micelles [2–5]. If cholesterol is replaced by the plant sterols in the micelles, it cannot be taken into the cells and be incorporated into chylomicrons.

Plant sterols block cholesterol absorption in the gut.

It has also been proposed that competition between cholesterol and plant sterols occurs for inclusion in the chylomicrons rather than micelles [5,6]. This small difference simply means that competition occurs at a later stage but the end point of reduced LDL-C production remains the same.

Additional Benefits

While the effects of plant sterols on cholesterol have been well studied, little research has been conducted examining whether their consumption confers any additional benefits to human health. Plant sterols may possess antiinflammatory [7] and antioxidation [8] properties, however further research needs to be conducted as to whether these properties will translate into meaningful health benefits. Additionally, a review by the European Atherosclerosis Society has concluded that 2g of plant sterols consumed per day significantly reduces LDL-C and may be considered for treatment of certain at risk individuals [9].

Health Claims and Plant Sterols

There are several health claims relating to plant sterols. Health claims are specific benefits that can be attributed to the consumption of certain foods. These claims are allowed to be displayed upon food packages to advertise their benefits to consumers. Each health claim is regulated by law and based on an extensive review of scientific evidence in most jurisdictions.

A health claim exists for plant sterols' role in reducing cholesterol.

The addition of plant sterols to a specific range of foods has been approved by the US Food and Drug Administration since 2000, Health Canada's Food Directorate since 2010 and the European Food Safety Authority since 2014 following a complete safety assessment. Health Canada also approved a cholesterol reduction health claim for plant sterols. At the same time, the body of evidence demonstrating the LDL-C lowering effects of plant sterols was reviewed. This led to the granting of a health claim for plant sterol-enriched products (margarine, mayonnaise, yogurt, salad dressing, and fruit and vegetable drinks) that contained at least 0.65 g/serving of plant sterols, among other heart-healthy ingredient requirements.

FIBER

Background

Simply put, fiber is any carbohydrate found in plants that is indigestible in the small intestine, although some can be fermented and broken down in the colon. There are two main types of fiber: soluble and insoluble (Fig. 3.2). Soluble fiber dissolves easily in water, while insoluble fiber does not. Think of what happens when you add hot water to oats, a food full of soluble fiber. The thick sticky porridge that results is because of the soluble fiber the oats contain. Now imagine what happens when you pour water over wheat bran. It does not absorb much water at all. This result is because wheat bran is primarily composed of insoluble fiber. As insoluble fiber does not dissolve in water it provides bulking to the stool and acts to prevent constipation. Soluble fiber acts to soften stool through absorption of water and may be beneficial in the treatment of both diarrhea and constipation.

Soluble fiber can be viscous and non viscous (Fig. 3.2).

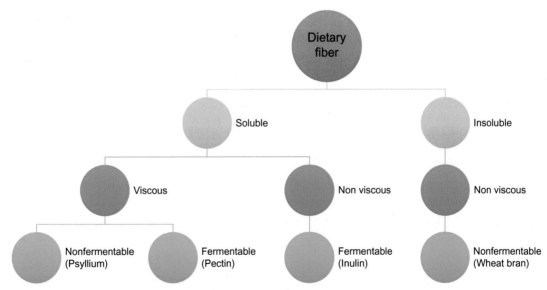

FIG. 3.2   Dietary fiber classifications.

Viscous fiber acts as a thickener in the intestinal tract forming a gel that slows stomach contents on its way to the small intestine and reduces the rate of carbohydrate absorption. Some viscous fibers can be partially or fully fermented in the colon like barley and oats, which may provide benefits to the microbiome, while others are minimally or not fermented at all, like psyllium. In the Portfolio diet, both types of viscous fiber were included, the main sources being: psyllium, okra, eggplant, oats, barley, persimmons and some pulses and legumes. When 20g of viscous fiber per day was added to the Portfolio diet from these sources, significant reductions in LDL-C were observed.

Mechanisms

Classifying fiber based on its viscosity (viscous and non viscous) rather than its solubility (soluble and insoluble), may in fact be a more useful distinction. The reason being that viscosity accounts for the ability of specific soluble fibers to reduce blood glucose and cholesterol levels [10].

The ability of viscous fiber to reduce blood glucose and cholesterol can in part be attributed to the way in which this type of fiber behaves in the intestinal tract, namely resisting flow and forming a gel which delays the absorption of nutrients like glucose (Fig. 3.3). The slower release of glucose into the blood stream results in a lower insulin response [10,11], which has been shown to potentially reduce cholesterol production [12]. Viscous fiber slows absorption by increasing the thickness of the unstirred water layer. The unstirred water layer is situated between the absorption cell (enterocytes) and the inner space of the small intestine (also known as the intestinal lumen) and forms a barrier which nutrients must first cross to be absorbed. In the unstirred water layer are the villi, long and noodle shaped structures which absorb nutrients. Their shape maximizes the surface area available to absorb the nutrients entering the small intestine. Once absorbed, the nutrients pass through the enterocyte cells, which are cells lining the intestine. In the enterocytes, nutrients are packaged and released into the blood stream or lymphocytes. If the unstirred water layer is thicker and offers more resistance to nutrients, they are not absorbed and released into the blood stream as quickly.

In addition to the thickening of the unstirred water layer, viscous fiber also creates a sticky gel in the lumen, which can trap nutrients and delay their absorption [13]. More specifically, the sticky gel that viscous fiber forms may function to trap bile acids, which are made from cholesterol, as seen in Fig. 3.4 [14,15]. Bile acids are usually reabsorbed by the body and reused. When they are trapped in the stool and subsequently excreted by the body, new cholesterol will have to be pulled from circulation and used to make more bile acids [16–18]. In this way circulating cholesterol is reduced.

> Viscous fiber slows carbohydrate absorption.

> Slower carbohydrate absorption, reduces the insulin response.

> Viscous fiber increases bile acid loss.

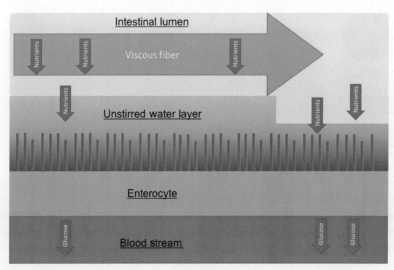

FIG. 3.3   Mechanism by which viscous fiber slows glucose absorption.

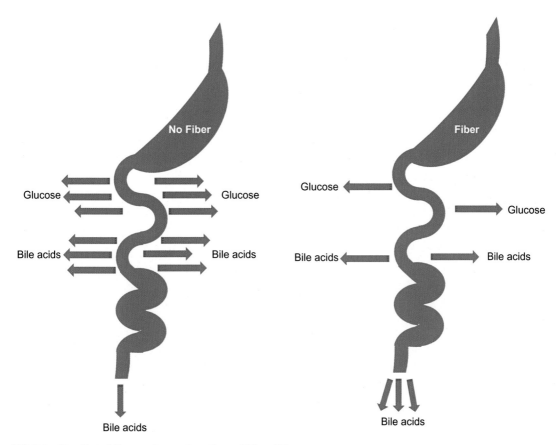

FIG. 3.4 The effect of fiber on glucose absorption and bile acid loss.

The presence of viscous fiber in the small intestine may interfere with micelle formation. Micelle formation allows fat and cholesterol to enter the cells lining the intestinal wall. By interfering with this process, viscous fiber is believed to prevent the absorption of fats and cholesterol [19–21]. Additionally, viscous fiber is fermented by colonic bacteria producing short-chain fatty acid (SCFA) [22]. Specific SCFAs may interfere with endogenous cholesterol production [1].

Additional Benefits

Soluble fiber has been found to delay gastric emptying [23], which can promote feelings of fullness. Because of this extended period of fullness, soluble fiber may be a useful tool for weight loss [13]. SCFA associated with soluble fiber have also seen to have benefits in nourishing the colonic mucosa and aiding in the prevention of colon cancer [24].

Psyllium has been found to be helpful for individuals suffering from irritable bowel syndrome, hemorrhoids, or constipation as it is a very gentle stool softener [25]. Conversely, it has also been shown to aid in the treatment of diarrhea and has been successful in managing extreme cases such as diarrhea induced by chemotherapy [26]. As mentioned above, viscous fibers can slow the release of glucose from the small intestine [10,11], which has implications for diabetic patients.

Viscous fiber can help keep you feeling full.

Health Claims and Soluble Fiber

There are several health claims relating to soluble fiber allowed by the Canadian and US governments to describe the health benefits to consumers.

Psyllium

In the United States, the FDA approves psyllium for reducing risk of coronary heart disease (CHD) [27]. Health Canada additionally approves psyllium for laxation or regularity of bowl movements stating that products containing 3.5 g or more can aid in this process [28].

Oats and Barley

In the United States, soluble fiber from whole oats is approved for a health claim stating they reduce the risk of CHD [27]. Beta-glucan from barley is approved in Canada for a health claim stating that it reduces blood cholesterol levels [28]. The viscous fiber beta-glucan found in oats is also approved for a health claim in Canada stating that it reduces postprandial glucose levels [28].

NUTS AND SEEDS

Background

What constitutes a nut varies greatly depending on the definition that is used. The culinary definition is the most inclusive, encompassing any shelled oily kernels that can be eaten. This allows legumes, such as the peanut, which actually grow underground, to be included in this category. Nuts are highly nutritious, often high in MUFA, plant sterol rich oil, and contain vegetable proteins, fiber, and phytochemicals (biologically active compounds in plants which often convey health benefits) (Fig. 3.5). Despite these attributes only a small selection of nuts have been tested for CVD-related health benefits. These nuts include peanuts, almonds, hazelnuts, and walnuts [29,30]. When 45 g of nuts per day were added to the Portfolio diet HDL-C was found to increase. Furthermore nuts, in general, lower serum LDL-C [31–33]. If allergies prevent the consumption of nuts, seeds can be substituted for nuts, as they are also high in MUFA, plant sterol rich oil, vegetable proteins, fiber, and phytochemicals and are believed to have similar effects [34].

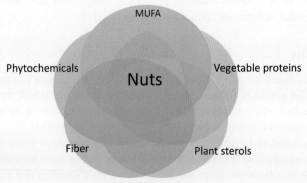

FIG. 3.5 Active components of nuts.

Mechanisms

There are many active components in nuts that work together to reduce CVD risk factors. To a large extent benefits can be attributed to the presence of MUFA (Box 3.1).

BOX 3.1 Monounsaturated Fatty Acids

MUFA are essentially fat molecules that contain one (mono) unsaturated or double bonded, carbon. This structure gives the molecule unique properties and causes it to interact differently from other fats in the body. MUFA is in a variety of foods including nuts but also oils. Typically oils that are liquid at room temperature and solid when chilled, such as olive oil and other plant-based fats, are high in MUFA as well.

Monounsaturated Fatty Acids

High-Density Lipoprotein

The MUFA in nuts may be responsible for the increase in HDL-C found after their addition to the Portfolio diet. One possible mechanism through which nuts may increase HDL-C is because they are used to replace calories usually consumed as carbohydrates (Fig. 3.6).

> Nuts decrease VLDL production and increase HDL-C.

The reduction in the carbohydrate intake works to increase HDL-C by first reducing the need for the liver to create VLDL to package the triglycerides produced from breaking down carbohydrate [35]. Secondly, when VLDL is made by the liver it uses cholesterol ester, so the reduction in VLDL synthesis means there is now an increased availability of cholesterol ester for incorporation into HDL-C molecules [36,37]. When HDL-C molecules have a higher amount of cholesterol ester they are less rapidly broken down by the enzyme lecithin cholesterol acyltransferase. This means that the HDL-C molecule is able to stay in circulation longer [38–40], effectively increasing HDL-C concentration. The longer HDL-C is in circulation the more it circulates in the blood stream scavenging LDL-C, transporting it to the liver and eliminating it from the body.

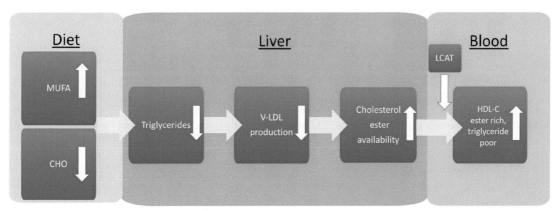

FIG. 3.6 Proposed pathway to HDL-C increase. *MUFA*, monounsaturated fatty acids, *CHO*, carbohydrates; *VDLD*, very low-density lipoprotein; *LCAT*, lecithin cholesterol acyltransferase; *HDL-C*, high-density lipoprotein cholesterol.

Apolipoprotein A1 and C-Reactive Protein

MUFA may decrease inflammatory biomarkers like C-reactive protein which are associated with an increased risk of CVD (Fig. 3.7). They may also increase the production of antiinflammatory biomarkers like Apo A1, which are associated with a decreased risk of CVD (Fig. 3.7). The process is proposed to occur by:

> Nuts decrease inflammation.

1. Reducing the production of proteins that stimulate inflammation in fat tissue. This action reduces the production of C-reactive protein by the liver. In addition, MUFA themselves are less susceptible to oxidization and are therefore less likely to release free radicals which can damage cells and create inflammation [41], as seen with polyunsaturated fatty acids.
2. The environment created by low levels of inflammation fosters the increased production of antiinflammatory proteins such as apolipoprotein A1.

Phytochemicals

Phytochemicals found in nuts have been suggested to contribute to the prevention of CVD by acting as antioxidants which prevent cell damage by scavenging free radicals and increasing the ability of arteries to adjust blood flow [42,43] (Box 3.2).

> In nuts, especially almonds, phytochemicals are predominately in the soft outer skin.

It is important to note that in nuts, especially almonds, phytochemicals are predominately in the soft outer skin. If it is removed before consumption, almost 50% of the antioxidants are lost [44,45].

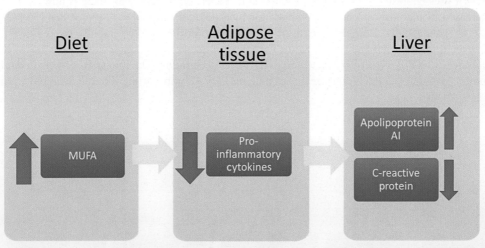

FIG. 3.7 Proposed monounsaturated fatty acid (MUFA) pathway to lowering C-reactive protein and increasing apolipoprotein A1.

BOX 3.2 Phytochemicals

Phytochemicals get their name from the Greek word phyton meaning plant. As the name suggests they are derived from plants and are considered nonnutritive or nonessential meaning we do not need them to survive. Phytochemicals are used to aid plants in fighting pathogens, predators and increasing their competitive ability. Many health benefits have been attributed to their consumption, making it apparent they are not only beneficial to plants.

Additional Benefits

Adding nuts to the diet has also been found to increase glycemic control for those with type 2 diabetes [46]. Nuts have shown to promote better fasting blood glucose levels, which is beneficial for people with or prone to diabetes . In addition they have been found to lower triglyceride levels which is beneficial for reducing the risk of CHD [47]. Waist circumference is also affected by increased nut consumption. When almonds were substituted for a calorie equivalent carbohydrate-based snack, individuals consuming almonds significantly decreased their waist circumference compared to those that consumed the same amount of calories in carbohydrates [48].

Nuts may improve glycemic control, and reduce waist circumference.

Health Claims and MUFA

While no specific health claims for MUFA exist, the majority of governing bodies and regulatory agencies agree that a diet that substitutes saturated or trans fats for unsaturated fat reduces cholesterol and the risk of heart disease. This claim has been recognized by the American Food and Drug Administration [49], the Canadian Food Inspection Agency [50], and the European Union [51]. Interestingly, in a recent analysis of 2 cohort studies, higher intake of MUFA derived from plant sources was associated with lower total mortality when compared to MUFA intake derived from animal products [52].

PLANT-BASED PROTEIN

Background

Plant-based proteins such as soybeans, chickpeas, lentils and other pulses are recognized for being a high-quality source of protein while also being low in saturated fat and high in PUFA, MUFA and in fiber [53] (Fig. 3.8). In the original Portfolio diet studies, soybeans were the main source of plant-based protein and were incorporated in the form of soy milk, soy meat analogs, and more traditional forms such as tofu and tempeh. These soy products were used as a substitute for cow's milk and meat-based protein sources.

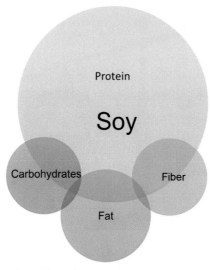

FIG. 3.8 Relative macronutrient composition of the soybean.

Soy protein was added to the original Portfolio diet studies as it had previously been found to reduce LDL-C [54,55]. When 45 g soy protein was consumed per day in the context of the Portfolio diet, reductions in LDL-C were observed. In the latest version of the Portfolio diet, the "Enhanced Portfolio", pulses such as chickpeas, lentils, beans and peas are also included as alternative protein sources. This is a direct result of the growing body of evidence suggesting they are also able to lower LDL-C [56].

Mechanism

Soy

Much of the research on the LDL-C-lowering properties of soy has centred on the bioactive peptides, otherwise known as protein building blocks of soy. These bioactive peptides have been suggested to reduce LDL-C through interference with LDL-C receptors, as well as bile acid regulation [57,58]. The reductions in LDL-C through these intrinsic mechanisms have similar potential to other Portfolio components, with LDL-C reductions ranging from ~4% to 6% [59–61]. Pulses contain many compounds that may contribute to their LDL-C lowering ability including viscous fiber, and polyphenols [62,63]. The viscous fiber found in pulses may reduce LDL-C by a number of mechanisms including increasing cholesterol loss through bile acid/salt excretion, slowing the absorption of glucose and consequently reduce insulin levels and increasing the production of beneficial SCFA through fermentation in the colon [1,22,64,65] (See Fiber section for more details).

Additionally, the polyphenols commonly found in pulses include carotenoids, phenolic acids, and tocopherols [64,66]. These are substances that may help prevent CVD, by protecting LDL-C from free radical oxidation [64,66].

Another significant benefit of soy along with other pulses may lie in what they replace in the diet. In other words, soy and other pulses can fill the role of less healthy, foods. Due to their high protein content soy and other pulses makes a great replacement food for red and processed meat. Soy and other pulse based foods, as mentioned above, are also low in fat and absent of cholesterol [67]. The traditional animal-based food products that soy and other pulses can replace are higher in saturated fat and contain cholesterol (Fig. 3.9). In the case of soy, researchers have investigated how much of a reduction in LDL-C

> Bioactive peptides in soy contribute to LDL-C reduction.

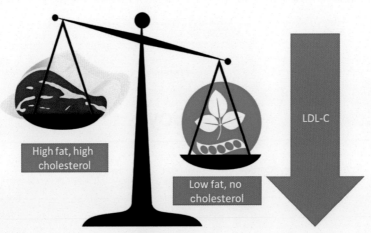

FIG. 3.9 The dietary replacement of meat by soy products.

can be observed based solely on what it would replace [68,69]. Using predictive equations, the displacement value of soy was calculated. It was found that replacing animal-based products with soy resulted in 3.6%–6% reduction in LDL-C. When the intrinsic value of soy (the 4%–6% LDL-C reduction) was overlaid with the above data, a combined potential reduction of 7.9%–10.3% in LDL-C was predicted. These reductions are important, it is rare for foods to reduce cholesterol more than 4% when added to the diet.

Additional Benefits

In addition to LDL-C reduction, the consumption of fiber found in soy and pulses has been related to improved weight-loss [70,71]. Furthermore, improvements in glycemic control, plasma lipid profiles, and reductions in blood pressure, have also been observed with the addition of soy polysaccharides (a carbohydrate found in soy) to the diet, which have viscous fiber properties [72]. Consumption of pulses has also shown to improve markers for longer term glycemic control [73]. Connections with soy and reduced risk of reproductive cancers and easement of menopause symptoms such as hot flashes have also been reported [75–79]. Soybeans also contain significant amounts of phytochemicals. These phytochemicals like isoflavins, phytosterols, and lecithins, are hypothesized to also exert cardio-protective effects [74] and are just now being examined in greater detail.

A health claim exists for soy's role in reducing CHD risk.

Health Claims and Soy

Taken together, this body of evidence has prompted the Canadian Government to issue a health claim for soy, stating that the consumption of 25 g/day can reduce the risk of CHD [80] (Box 3.3).

BOX 3.3 Soy and Health Concerns

Soy an history of use

Soy products such as tofu have been consumed for 1000's of years, particularly in East and South East Asia [81].

Groups such as the Buddhists monks have been hypothesized to perpetuate the spread of tofu owing to its usefulness in providing protein in a plant-based diet [81]. However, many other nonvegetarian groups also consumed tofu. The first written record of its consumption was by a Shinto priest in 1183 CE. Tofu was known throughout China and by 618–907 CE was commonly referred to as "small mutton" [81]. Tofu also spread to Japan where it became a widely appreciated art form. In the same way the French, notorious for bread making, critique a baguette from shop to shop, tofu undergoes the same scrutiny in Japan. Other regions, such as Indonesia, also adopted soy protein, but favored a variant called tempeh, a fermented soybean food [81].

Estrogen

One of the main concerns around soy consumption is a belief that soy promotes detrimental effects through increased estrogen activity. This belief stems from a fear of a group of compounds soy contains called phytoestrogens, which are very similar to estrogen. While these molecules may be similar to estrogen they behave very differently in the body and these health concerns have been shown to be unfounded. Unlike estrogen, phytoestrogens found in soy actually appear to reduce the risk of reproductive cancers such as breast, endometrial, and prostate cancer [75–78].

Continued

BOX 3.3 Soy and Health Concerns—Cont'd

Safety through history of use

Soy products such as tofu have been consumed for about 2000 years, particularly in East and South East Asia [81]. Groups such as the Buddhists monks have been hypothesized to have promoted the spread of tofu owing to its usefulness in providing protein in a plant-based diet [81]. However, many other nonvegetarian groups also consumed tofu. The first written record of its consumption was by a Shinto priest in 1183 CE. Tofu was known throughout China and by 618-907 CE was commonly referred to as "small mutton" [81]. Tofu also spread to Japan where it became a widely appreciated art form. In the same way the French, famous for bread making, critique a baguette from shop to shop, tofu undergoes the same scrutiny in Japan. Other regions, such as Indonesia, also adopted soy protein, but favored a variant called tempeh, a fermented soybean food [81]. The ancient and prolonged use of soy products such as tofu and tempeh stand as a testament to their nonharmful health effects.

Modern soy

An additional concern surrounds the modern methods of soy production and processing which differ from traditional practises. One of these concerns relates to the extent to which soy is genetically modified. In North America soy beans are one of the most commonly genetically modified (GM) crops. However, these soy beans are primarily used for animal feed [82]. While there is no evidence to suggest GM soy is bad for human or animal health, if it is a personal concern then buying organic products is always an option. Another modern change to soy products occurs at the level of processing. A chemical called hexane is sometimes used to extract oil from foods like soybeans. For soy, this extraction is done to create meat analogs and protein powders and is listed on the label as "soy protein isolate," "textured soy protein," or "soy protein concentrate." While hexane is approved for use in food by the Food and Drug administration (FDA), according to the Center for Disease Control (CDC), high levels of hexane are considered neurotoxic and are classified as an air pollutant by the EPA. The extraction process therefore potentially increases the environmental impact of these products [83–85]. Although most evidence indicates that hexane is predominately removed after processing, in order to avoid potential of contamination, concerned individuals can try switching to organic products which have banned the use of hexane. Alternatively, consider consuming more products using whole soy beans such as tofu, tempeh, soy yogurts, etc.

REFERENCES

[1] Slavin J. Fiber and prebiotics: mechanisms and health benefits. Nutrients 2013;5(4):1417–35. https://doi.org/10.3390/nu5041417.

[2] Armstrong MJ, Carey MC. Thermodynamic and molecular determinants of sterol solubilities in bile salt micelles. J Lipid Res 1987;28(10):1144–55. Retrieved from, http://www.ncbi.nlm.nih.gov/pubmed/3681139.

[3] Brown AW, Hang J, Dussault PH, Carr TP. Phytosterol ester constituents affect micellar cholesterol solubility in model bile. Lipids 2010;45(9):855–62. https://doi.org/10.1007/s11745-010-3456-6.

[4] Klett EL, Lu K, Kosters A, Vink E, Lee M-H, Altenburg M, et al. A mouse model of sitosterolemia: absence of Abcg8/sterolin-2 results in failure to secrete biliary cholesterol. BMC Med 2004;2(1):5. https://doi.org/10.1186/1741-7015-2-5.

[5] Salen G, Ahrens EH, Grundy SM. Metabolism of β-sitosterol in man. J Clin Investig 1970;49(5):952–67. https://doi.org/10.1172/JCI106315.

[6] Field FJ, Born E, Mathur SN. Effect of micellar beta-sitosterol on cholesterol metabolism in CaCo-2 cells. J Lipid Res 1997;38(2):348–60. Retrieved from, http://www.ncbi.nlm.nih.gov/pubmed/9162754.

[7] Bouic PJ. The role of phytosterols and phytosterolins in immune modulation: a review of the past 10 years. Curr Opin Clin Nutr Metab Care 2001;4:471–5. https://doi.org/10.1097/00075197-200111000-00001.

[8] van Rensburg SJ, Daniels WM, van Zyl JM, Taljaard JJ. A comparative study of the effects of cholesterol, beta-sitosterol, beta-sitosterol glucoside, dehydroepiandrosterone sulphate and melatonin on in vitro lipid peroxidation. Metab Brain Dis 2000;15:257–65. https://doi.org/10.1023/A:1011167023695.

[9] Gylling H, Plat J, Turley S, Ginsberg HN, Ellegård L, Jessup W, et al. European Atherosclerosis Society Consensus Panel on Phytosterols. Plant sterols and plant stanols in the management of dyslipidaemia and prevention of cardiovascular disease. Atherosclerosis 2014;232(2):346–60. https://doi.org/10.1016/j.atherosclerosis.2013.11.043.

[10] Jenkins DJ, Wolever TM, Leeds AR, Gassull MA, Haisman P, Dilawari J, et al. Dietary fibres, fibre analogues, and glucose tolerance: importance of viscosity. Br Med J 1978;1(6124):1392–4. Retrieved from, http://www.ncbi.nlm.nih.gov/pubmed/647304.

[11] Bourdon I, Yokoyama W, Davis P, Hudson C, Backus R, Richter D, et al. Postprandial lipid, glucose, insulin, and cholecystokinin responses in men fed barley pasta enriched with beta-glucan. Am J Clin Nutr 1999;69(1):55–63. Retrieved from, http://www.ncbi.nlm.nih.gov/pubmed/9925123.

[12] Sample CE, Ness GC. Regulation of the activity of 3-hydroxy-3-methylglutaryl coenzyme A reductase by insulin. Biochem Biophys Res Commun 1986;137(1):201–7. Retrieved from http://www.ncbi.nlm.nih.gov/pubmed/3521603.

[13] Howarth NC, Saltzman E, Roberts SB. Dietary fiber and weight regulation. Nutr Rev 2001;59(5):129–39. Retrieved from, http://www.ncbi.nlm.nih.gov/pubmed/11396693.

[14] Kritchevsky D, Story JA. Binding of bile salts in vitro by nonnutritive fiber. J Nutr 1974;104(4):458–62. Retrieved from, http://www.ncbi.nlm.nih.gov/pubmed/4816931.

[15] Kritchevsky D, Story JA. Letter: In vitro binding of bile acids and bile salts. Am J Clin Nutr 1975;28(4):305–6. Retrieved from, http://www.ncbi.nlm.nih.gov/pubmed/1119427.

[16] Gelissen IC, Brodie B, Eastwood MA. Effect of Plantago ovata (psyllium) husk and seeds on sterol metabolism: studies in normal and ileostomy subjects. Am J Clin Nutr 1994;59(2):395–400. Retrieved from, http://www.ncbi.nlm.nih.gov/pubmed/8310991.

[17] Jenkins DJA, Kendall CWC, Jackson C-JC, Connelly PW, Parker T, Faulkner D, et al. Effects of high- and low-isoflavone soyfoods on blood lipids, oxidized LDL, homocysteine, and blood pressure in hyperlipidemic men and women. Am J Clin Nutr 2002;76(2):365–72.

[18] Jenkins DJ, Wolever TM, Rao AV, Hegele RA, Mitchell SJ, Ransom TP, et al. Effect on blood lipids of very high intakes of fiber in diets low in saturated fat and cholesterol. N Engl J Med 1993;329(1):21–6. https://doi.org/10.1056/NEJM199307013290104.

[19] Glore SR, Van Treeck D, Knehans AW, Guild M. Soluble fiber and serum lipids: a literature review. J Am Diet Assoc 1994;94(4):425–36. Retrieved from, http://www.ncbi.nlm.nih.gov/pubmed/8144811.

[20] Haskell WL, Spiller GA, Jensen CD, Ellis BK, Gates JE. Role of water-soluble dietary fiber in the management of elevated plasma cholesterol in healthy subjects. Am J Cardiol 1992;69(5):433–9. Retrieved from, http://www.ncbi.nlm.nih.gov/pubmed/1310566.

[21] Vahouny GV, Tombes R, Cassidy MM, Kritchevsky D, Gallo LL. Dietary fibers: V. Binding of bile salts, phospholipids and cholesterol from mixed micelles by bile acid sequestrants and dietary fibers. Lipids 1980;15(12):1012–8. Retrieved from, http://www.ncbi.nlm.nih.gov/pubmed/6261073.

[22] Gunness P, Gidley MJ. Mechanisms underlying the cholesterol-lowering properties of soluble dietary fibre polysaccharides. Food Funct 2010;1(2):149. https://doi.org/10.1039/c0fo00080a.

[23] Blackburn NA, Redfern JS, Jarjis H, Holgate AM, Hanning I, Scarpello JH, et al. The mechanism of action of guar gum in improving glucose tolerance in man. Clin Sci (London, England: 1979) 1984;66(3):329–36. Retrieved from http://www.ncbi.nlm.nih.gov/pubmed/6362961.

[24] Wong, Julia de Souza R, Kendall C, Emam A, Jenkins D. Colonic Health: Fermentation and Short Chain Fatty Acids: Journal of Clinical Gastroenterology. Retrieved from https://journals.lww.com/jcge/Abstract/2006/03000/Colonic_Health__Fermentation_and_Short_Chain_Fatty.15.aspx; 2006.

[25] Rigaud D, Paycha F, Meulemans A, Merrouche M, Mignon M. Effect of psyllium on gastric emptying, hunger feeling and food intake in normal volunteers: a double blind study. Eur J Clin Nutr 1998;52(4):239–45. Retrieved from, http://www.ncbi.nlm.nih.gov/pubmed/9578335.

[26] Muehlbauer PM, Thorpe D, Davis A, Drabot R, Rawlings BL, Kiker E. Putting evidence into practice: evidence-based interventions to prevent, manage, and treat chemotherapy-and radiotherapy-induced diarrhea. Clin J Oncol Nurs 2009;13(3). https://doi.org/10.1188/09.CJON.336-341.

[27] FDA Department of Health and Human Services. CFR-Code of Federal Regulations Title 21: 173.255 Methylene chloride. Retrieved from, https://www.accessdata.fda.gov/scripts/cdrh/cfdocs/cfcfr/cfrsearch.cfm?fr=101.81; 2017.

[28] Health Canada List of Dietary Fibres Reviewed and Accepted by Health Canada's Food Directorate-Canada.ca. March 11, 2018. Retrieved from https://www.canada.ca/en/health-canada/services/publications/food-nutrition/list-reviewed-accepted-dietary-fibres.html.

[29] Jenkins DJA, Jones PJH, Lamarche B, Kendall CWC, Faulkner D, Cermakova L, et al. Effect of a dietary portfolio of cholesterol-lowering foods given at 2 levels of intensity of dietary advice on serum lipids in hyperlipidemia: a randomized controlled trial. JAMA 2011;306(8):831–9. https://doi.org/10.1001/jama.2011.1202.

[30] Njike VY, Costales VC, Petraro P, Annam R, Yarandi N, Katz DL. The resulting variation in nutrient intake with the inclusion of walnuts in the diets of adults at risk for type 2 diabetes: a randomized, controlled, crossover trial. Am J Health Promot. 2019;33(3):430–8. https://doi.org/10.1177/0890117118791120.

[31] Estruch R, Ros E, Salas-Salvadó J, Covas MI, Corella D, Arós F, et al. PREDIMED Study Investigators. Primary prevention of cardiovascular disease with a Mediterranean diet supplemented with extra-virgin olive oil or nuts. N Engl J Med 2018;378(25):e34. https://doi.org/10.1056/NEJMoa1800389.

[32] Sabaté J, Oda K, Ros E. Nut consumption and blood lipid levels: a pooled analysis of 25 intervention trials. Arch Intern Med 2010;170(9):821–7. https://doi.org/10.1001/archinternmed.2010.79.

[33] Jenkins DJ, Kendall CW, Marchie A, Parker TL, Connelly PW, Qian W, et al. Dose response of almonds on coronary heart disease risk factors: blood lipids, oxidized low-density lipoproteins, lipoprotein(a), homocysteine, and pulmonary nitric oxide: a randomized, controlled, crossover trial. Circulation 2002;106(11):1327–32.

[34] Chiavaroli L, Nishi SK, Khan TA, Braunstein CR, Glenn AJ, Mejia SB, et al. Portfolio dietary pattern and cardiovascular disease: a systematic review and meta-analysis of controlled trials. Prog Cardiovasc Dis 2018;61(1):43–53. https://doi.org/10.1016/j.pcad.2018.05.004.

[35] Rashid S, Watanabe T, Sakaue T, Lewis GF. Mechanisms of HDL lowering in insulin resistant, hypertriglyceridemic states: the combined effect of HDL triglyceride enrichment and elevated hepatic lipase activity. Clin Biochem 2003;36(6):421–9. Retrieved from, http://www.ncbi.nlm.nih.gov/pubmed/12951168.

[36] Lewis GF. Determinants of plasma HDL concentrations and reverse cholesterol transport. Curr Opin Cardiol 2006;21(4):345–52. https://doi.org/10.1097/01.hco.0000231405.76930.a0.

[37] Tall AR. Cholesterol efflux pathways and other potential mechanisms involved in the athero-protective effect of high density lipoproteins. J Intern Med 2008;263(3):256–73. https://doi.org/10.1111/j.1365-2796.2007.01898.x.

[38] Lewis GF, Rashid S, Uffelman KD, Lamarche B. Mechanism of HDL Lowering in Insulin Resistant States. Boston, MA: Springer; 2001. p. 273–7. https://doi.org/10.1007/978-1-4615-1321-6_34.

[39] Nälsén C, Vessby B, Berglund L, Uusitupa M, Hermansen K, Riccardi G, et al. Dietary (n-3) fatty acids reduce plasma F2-isoprostanes but not prostaglandin F2alpha in healthy humans. J Nutr 2006;136(5):1222–8. Retrieved from, http://www.ncbi.nlm.nih.gov/pubmed/16614408.

[40] Tall AR. Plasma high density lipoproteins. Metabolism and relationship to atherogenesis. J Clin Investig 1990;86(2):379–84. https://doi.org/10.1172/JCI114722.

[41] Reaven PD, Witztum JL. Oxidized low density lipoproteins in atherogenesis: role of dietary modification. Annu Rev Nutr 1996;16(1):51–71. https://doi.org/10.1146/annurev.nu.16.070196.000411.

[42] Kris-Etherton PM, Hu FB, Ros E, Sabaté J. The role of tree nuts and peanuts in the prevention of coronary heart disease: multiple potential mechanisms. J Nutr 2008;138(9):1746S–51S. https://doi.org/10.1016/j.mad.2013.11.011.

[43] Njike VY, Yarandi N, Petraro P, Ayettey RG, Treu JA, Katz DL. Inclusion of walnut in the diets of adults at risk for type 2 diabetes and their dietary pattern changes: a randomized, controlled, cross-over trial. BMJ Open Diabetes Res Care 2016;4(1):e000293. https://doi.org/10.1136/bmjdrc-2016-000293.

[44] Blomhoff R, Carlsen MH, Andersen LF, Jacobs DR. Health benefits of nuts: potential role of antioxidants. Br J Nutr 2006;96(Suppl. 2):S52–60. Retrieved from, http://www.ncbi.nlm.nih.gov/pubmed/17125534.

[45] Chen C-Y, Milbury PE, Lapsley K, Blumberg JB. Flavonoids from almond skins are bioavailable and act synergistically with vitamins C and E to enhance hamster and human LDL resistance to oxidation. J Nutr 2005;135(6):1366–73.

[46] Viguiliouk E, Kendall CWC, Blanco Mejia S, Cozma AI, Ha V, Mirrahimi A, et al. Effect of tree nuts on glycemic control in diabetes: a systematic review and meta-analysis of randomized controlled dietary trials. PLoS One 2014;9(7):e103376. https://doi.org/10.1371/journal.pone.0103376.

[47] Blanco Mejia S, Kendall CWC, Viguiliouk E, Augustin LS, Ha V, Cozma AI, et al. Effect of tree nuts on metabolic syndrome criteria: a systematic review and meta-analysis of randomised controlled trials. BMJ Open 2014;4(7):e004660. https://doi.org/10.1136/bmjopen-2013-004660.

[48] Berryman CE, West SG, Fleming JA, Bordi PL, Kris-Etherton PM. Effects of daily almond consumption on cardiometabolic risk and abdominal adiposity in healthy adults with elevated LDL-cholesterol: a randomized controlled trial. J Am Heart Assoc 2015;4(1):e000993. https://doi.org/10.1161/JAHA.114.000993.

[49] Canadian Food Inspection Agency. Acceptable disease risk reduction claims and therapeutic claims – Health Claims – Food. Canadian Food Inspection Agency; 2018. Retrieved from: http://www.inspection.gc.ca/food/general-food-requirements-and-guidance/labelling/for-industry/health-claims/eng/1392834838383/1392834887794?chap=7#s8c7.

[50] Food and Drug Administration. Labeling and nutrition – Health claim notification for the substitution of saturated fat in the diet with unsaturated fatty acids and reduced risk of heart disease. Center for Food Safety and Applied Nutrition; 2007. Retrieved from: https://www.fda.gov/Food/LabelingNutrition/ucm073631.htm.

[51] Livingstone KM. Authorised EU health claim for MUFA and PUFA in replacement of saturated fats. In: Foods, nutrients and food ingredients with authorised EU health claims. 2018. p. 87–100. https://doi.org/10.1016/B978-0-08-100922-2.00006-1.

[52] Guasch-Ferré M, Zong G, Willett WC, Zock PL, Wanders AJ, Hu FB, et al. Associations of monounsaturated fatty acids from plant and animal sources with total and cause-specific mortality in two US prospective cohort studies. Circ Res 2019;124(8):1266–75. https://doi.org/10.1161/CIRCRESAHA.118.313996.

[53] Katz DL, Doughty KN, Geagan K, Jenkins DA, Gardner CD. Perspective: the public health case for modernizing the definition of protein quality. Adv Nutr 2019; pii:nmz023. https://doi.org/10.1093/advances/nmz023.

[54] Anderson JW, Johnstone BM, Cook-Newell ME. Meta-analysis of the effects of soy protein intake on serum lipids. N Engl J Med 1995;333(5):276–82. https://doi.org/10.1056/NEJM199508033330502.

[55] Sirtori CR, Agradi E, Conti F, Mantero O, Gatti E. Soybean-protein diet in the treatment of type-II hyperlipoproteinaemia. Lancet (London, England) 1977;1(8006):275–7.

[56] Ha V, Sievenpiper JL, de Souza RJ, Jayalath VH, Mirrahimi A, Agarwal A, et al. Effect of dietary pulse intake on established therapeutic lipid targets for cardiovascular risk reduction: a systematic review and meta-analysis of randomized controlled trials. Can Med Assoc J 2014;186(8):E252–62. https://doi.org/10.1503/cmaj.131727.

[57] Maki KC, Butteiger DN, Rains TM, Lawless A, Reeves MS, Schasteen C, Krul ES. Effects of soy protein on lipoprotein lipids and fecal bile acid excretion in men and women with moderate hypercholesterolemia. J Clin Lipidol 2010;4(6):531–42. https://doi.org/10.1016/j.jacl.2010.09.001.

[58] Torres N, Torre-Villalvazo I, Tovar AR. Regulation of lipid metabolism by soy protein and its implication in diseases mediated by lipid disorders. J Nutr Biochem 2006;17(6):365–73. https://doi.org/10.1016/j.jnutbio.2005.11.005.

[59] Harland JI, Haffner TA. Systematic review, meta-analysis and regression of randomised controlled trials reporting an association between an intake of circa 25g soya protein per day and blood cholesterol. Atherosclerosis 2008;200(1):13–27. https://doi.org/10.1016/j.atherosclerosis.2008.04.006.

[60] Reynolds K, Chin A, Lees KA, Nguyen A, Bujnowski D, He J. A meta-analysis of the effect of soy protein supplementation on serum lipids. Am J Cardiol 2006;98(5):633–40. https://doi.org/10.1016/j.amjcard.2006.03.042.

[61] Zhan S, Ho SC. Meta-analysis of the effects of soy protein containing isoflavones on the lipid profile. Am J Clin Nutr 2005;81(2):397–408. Retrieved from, http://www.ncbi.nlm.nih.gov/pubmed/15699227.

[62] Padhi EMT, Ramdath DD. A review of the relationship between pulse consumption and reduction of cardiovascular disease risk factors. J Funct Foods 2017;38:635–43. https://doi.org/10.1016/j.jff.2017.03.043.

[63] Rochfort S, Panozzo J. Phytochemicals for health, the role of pulses. J Agric Food Chem 2007;55(20):7981–94. https://doi.org/10.1021/jf071704w.

[64] Bouchenak M, Lamri-Senhadji M. Nutritional quality of legumes, and their role in cardiometabolic risk prevention: a review. J Med Food 2013;16(3):185–98. https://doi.org/10.1089/jmf.2011.0238.

[65] Hutchins AM, Winham DM, Thompson SV. Phaseolus beans: impact on glycaemic response and chronic disease risk in human subjects. Br J Nutr 2012;108(S1):S52–65. https://doi.org/10.1017/S0007114512000761.

[66] Ros E, Hu FB. Consumption of plant seeds and cardiovascular health. Circulation 2013;128(5):553–65. https://doi.org/10.1161/CIRCULATIONAHA.112.001119.

[67] Rebello CJ, Greenway FL, Finley JW. Whole grains and pulses: a comparison of the nutritional and health benefits. J Agric Food Chem 2014;62(29):7029–49. https://doi.org/10.1021/jf500932z.

[68] Jenkins DJA, Chiavaroli L, Wong JMW, Kendall C, Lewis GF, Vidgen E, et al. Adding monounsaturated fatty acids to a dietary portfolio of cholesterol-lowering foods in hypercholesterolemia. CMAJ 2010;182(18):1961–7. https://doi.org/10.1503/cmaj.092128.

[69] Jenkins DJA, Mirrahimi A, Srichaikul K, Berryman CE, Wang L, Carleton A, et al. Soy protein reduces serum cholesterol by both intrinsic and food displacement mechanisms. J Nutr 2010;140(12):2302S–11S. https://doi.org/10.3945/jn.110.124958.

[70] Hu X, Gao J, Zhang Q, Fu Y, Li K, Zhu S, Li D. Soy fiber improves weight loss and lipid profile in overweight and obese adults: a randomized controlled trial. Mol Nutr Food Res 2013;57(12):2147–54. https://doi.org/10.1002/mnfr.201300159.

[71] Kim SJ, de Souza RJ, Choo VL, Ha V, Cozma AI, Chiavaroli L, et al. Effects of dietary pulse consumption on body weight: a systematic review and meta-analysis of randomized controlled trials. Am J Clin Nutr 2016;103(5):1213–23. https://doi.org/10.3945/ajcn.115.124677.

[72] Bouchenak M, Lamri-Senhadji M. Nutritional quality of legumes, and their role in cardiometabolic risk prevention: a review. J Med Food 2013;16(3):185–98. https://doi.org/10.1089/jmf.2011.0238.

[73] Sievenpiper JL, Kendall CWC, Esfahani A, Wong JMW, Carleton AJ, Jiang HY, et al. Effect of non-oil-seed pulses on glycaemic control: a systematic review and meta-analysis of randomised controlled experimental trials in people with and without diabetes. Diabetologia 2009;52(8):1479–95. https://doi.org/10.1007/s00125-009-1395-7.

[74] Ramdath DD, Padhi EMT, Sarfaraz S, Renwick S, Duncan AM. Beyond the cholesterol-lowering effect of soy protein: a review of the effects of dietary soy and its constituents on risk factors for cardiovascular disease. Nutrients 2017;9(4). https://doi.org/10.3390/nu9040324.

[75] Albuquerque RC, Baltar VT, Marchioni DM. Breast cancer and dietary patterns: a systematic review. Nutr Rev 2014;72(1):1–17. https://doi.org/10.1111/nure.12083.

[76] van Die MD, Bone KM, Williams SG, Pirotta MV. Soy and soy isoflavones in prostate cancer: a systematic review and meta-analysis of randomized controlled trials. BJU Int 2014;113(5b):E119–30. https://doi.org/10.1111/bju.12435.

[77] Wu J, Zeng R, Huang J, Li X, Zhang J, Ho J, Zheng Y. Dietary protein sources and incidence of breast cancer: a dose-response meta-analysis of prospective studies. Nutrients 2016;8(11):730. https://doi.org/10.3390/nu8110730.

[78] Zhang G-Q, Chen J-L, Liu Q, Zhang Y, Zeng H, Zhao Y. Soy intake is associated with lower endometrial cancer risk. Medicine 2015;94(50):e2281. https://doi.org/10.1097/MD.0000000000002281.

[79] Franco OH, Chowdhury R, Troup J, Voortman T, Kunutsor S, Kavousi M, et al. Use of plant-based therapies and menopausal symptoms. JAMA 2016;315(23):2554. https://doi.org/10.1001/jama.2016.8012.

[80] Health Canada. Summary of Health Canada's Assessment of a Health Claim about soy protein and cholesterol lowering. In: Bureau of Nutritional Sciences, Food Directorate, Health Products and Food Branch; 2015. p. 1–12. March, Retrieved from, https://www.canada.ca/en/health-canada/services/food-nutrition/food-labelling/health-claims/assessments/summary-assessment-health-claim-about-protein-cholesterol-lowering.html.

[81] Du Bois C: The story of soy. London: Reaktion Book; 2018.

[82] Food and Agriculture Organization. Genetically Modified Crops; 2014. p. 1–2. Retrieved from, www.fao.org/biotech/en/

[83] Center for Disease Control and Prevention. CDC-NIOSH 1988 OSHA PEL project documentation: list by chemical name: n-HEXANE; 1988.

[84] Environmental Protection Agency. Hexane; 2000. Summery Report. 110-54-3.

[85] US Food and Drug Administration. Food additives & ingredients-food additive status list. Center for Food Safety and Applied Nutrition, Silver Spring, MD; 2018.

Chapter 4

The Pantry: Tips and Tricks

INTRODUCTION

Starting a new diet can be a challenge. Although we have tried our best to limit the foods used in the Portfolio diet recipes to items easily found in most grocery stores, some components (like psyllium), may require a trip to the health food store. As mentioned previously, the Portfolio diet is recommended to be consumed within the context of a plant-based diet. This may mean the introduction of several unfamiliar foods. Knowing how to prepare these foods can present an additional challenge.

Although some dairy, meat, and egg substitutes come very close to replicating what they are imitating, in most cases they will not serve as exact copies. As you read through this chapter you will notice that these substitutes are often named after the foods they are meant to replace. For example, soy "milk." This naming system is simply to try and make it easier for people who wish to make plant-based versions of familiar recipes. While these products are meant to fill a similar niche as their animal product counterparts, they also stand up in their own right in terms of interesting flavor combinations and textures. Everyone has different preferences and it may take some experimentation to find the ones you like the most.

To accelerate this process we have complied a chapter that outlines different tips and tricks to help you start and stick to the Portfolio diet. In this chapter you will find an overview of the most commonly consumed (and possibly unfamiliar) foods that make up the bulk of the Dietary Portfolio as well as information on where to purchase these foods and how best to prepare them.

As always, before starting a new diet plan, you may wish to consult your physician, particularly if you hope to go off or reduce amounts of medication. Asking to be referred to a dietitian may also be helpful as they will be able to answer all of your diet-related questions.

STARTING UP

When you first begin to incorporate Portfolio foods into your diet, it may be beneficial to slowly increase the amounts you are eating. This is particularly true for viscous fiber, which can cause bloating and excessive gas if added too quickly to the diet. In order to avoid these side-effects try starting off with ¼ the recommended amount and see how your body reacts. It may take time for adjustment so feel free to play with amounts.

When you first begin to incorporate Portfolio foods into your diet, it may be beneficial to slowly increase the amounts you are eating.

The Portfolio Diet for Cardiovascular Disease Risk Reduction. https://doi.org/10.1016/B978-0-12-810510-8.00004-2

47

Your body will adapt over time. Changing eating patterns can be difficult, so do not be afraid to start slowly and adapt the recipes in this book to your comfort level. In Table 4.1, you will find a list of the most common ingredients used in the Portfolio diet and their most common location in a grocery or health food store. At the end of this chapter you will also find a 3-day eating guide (Table 4.3) to help kick start your meal planning and give you some inspiration as you begin your journey with the Portfolio diet.

TABLE 4.1   Shopping List

	Food	Location
High MUFA Oils	Olive oil	Grocery store • Baking aisle
	Canola oil	Grocery store • Baking aisle
	Sunflower oil	Grocery store • Baking aisle
Nuts & Seeds	Nuts (E.g. Almonds, walnuts, pecans, hazelnuts)	Bulk food store Grocery store • Baking aisle • Snack food aisle • Health food section
	Seeds (E.g. Sunflower, sesame, pumpkin, hemp)	Bulk food store Grocery store • Baking aisle • Snack food aisle • Health food section
Plant-Based Proteins	Legumes	Bulk food store (dried) Grocery store • Dried goods • Canned goods • Frozen section
	Soy Dairy Alternatives (E.g. Milk and yogurt)	Grocery store • Health food section • Dairy section
	Soy Mayonnaise	Grocery store • Health food section • Dairy section
	Soy Meat Alternatives (E.g. Texturized Vegetable Protein, tofu)	Grocery store • Health food section • Produce section cooler

Continued

TABLE 4.1 Shopping List—Cont'd

Plant Sterols	Sterol Enriched Juices (E.g. Orange juice)	Grocery store • Dairy section
	Sterol Enriched Margarine (E.g. Becel Proactiv®)	Grocery store • Dairy section
	Sterol Supplements	Grocery store • Health food section • Supplements aisle
Viscous Fiber	Barley	Bulk food store Grocery store • Dried goods
	Berries (E.g. Strawberries, blackberries, blueberries, raspberries)	Grocery store • Produce section • Frozen section
	Eggplant	Grocery store • Produce section
	Oats and Oat Bran	Bulk food store Grocery store • Cereal aisle
	Okra	Grocery store • Produce section • Frozen section
	Psyllium	Pharmacy (Metamucil®) Bulk store (husk) Grocery store • Supplements aisle (husk)

PORTFOLIO FOODS SERVING SIZES

The Dietary Portfolio consists of four key components; nuts and seeds, viscous fiber, plant-based protein, and plant sterols. For most foods the daily amount to be consumed varies based on the total number of calories you consume. The exception to this rule is berries (found under viscous fiber) 450g are recommended to be consumed daily regardless of calorie intake. Most adults need between 1800 and 2000 kcal per day. You can visit Health Canada to see exactly how many calories you need to consume each day based on your age, sex, and activity level at "Estimated Energy Requirements-Health Canada."[1] Table 4.2 shows the recommended daily intake of Portfolio foods per 1000 kcal. To find out how much you should be consuming divide your recommended calorie requirement by 1000 and multiply by the serving per day. For example, if you are recommended caloric intake is 1900 kcal you would consume 38 g of nuts per day.

---

[1] https://www.canada.ca/en/health-canada/services/food-nutrition/canada-food-guide/food-guide-basics/estimated-energy-requirements.html.

TABLE 4.2 Reccomened Daily Intake of Portfolio Foods

Portfolio Food	Serving per Day	Examples of Sources	Amounts and Sizes
Nuts and Seeds	20 g per 1000 kcal	Almonds, walnuts, pecans, hazelnuts, pumpkin seeds, sunflower seeds	A small handful—about 9–10 nuts equals 10 g
Plant-based protein	20 g per1000 kcal	Soy milk, soy meat analogs, tofu, edamame, tempeh, soy yogurt	1 cup soy milk = 8 g ½ cup edamame = 11 g 1 srv soy meat = 6 g 1 block tofu (size of your fist or 395 g) = 33 g 7 strips tempeh = 15 g 1 srv soy yogurt = 7 g
Plant sterols	1 g per 1000 kcal, with a minimum of 2 g per day and a maximum of 3 g per day	Supplements, sterol enriched margarine (e.g. Becel ProActive®), sterol enriched juices, avocados	2 tsp. Becel ProActiv® = 1 g Half an avocado = 0.8 g Tip: 1 tsp. is about the same size as the tip of your pinky while 1 tbsp. equals your thumb
Viscous Fiber	10 g per 1000 kcal	Oats, barley, psyllium, eggplant, okra	0.5 cup oat bran = 3 g 0.5 cup barley ckd = 1.5 g 1 tbsp. psyllium husk = 3 g 1 cup eggplant ckd = 1 g 1 cup okra ckd = 2 g
+	450 g Berries per day	Strawberries, blueberries, raspberries, blackberries	3 cups whole strawberries/blackberries 3 1/2 cups raspberries 4 1/2 cups blueberries

Note: *ckd*, cooked; *srv*, serving.

KEY PORTFOLIO FOODS: QUICK FACTS

Nuts and Seeds

Nuts are nutritious and delicious, great for cooking or eating on their own. They are a source of

- Vitamin E
- Magnesium
- Fiber
- Protein
- Monounsaturated and polyunsaturated fatty acids

Due to their high protein content, nuts have the added benefit of keeping you feeling full between meals and have shown benefits for reduction of abdominal fat. Nuts, like almonds, are beneficial for heart health and have been shown to lower cholesterol by 1% for every 10 g (~10 almonds) eaten per day. Nuts can be used for cooking and can be eaten raw, roasted, ground, slivered, sliced, blanched, or blended as nut butters. Look for nut butters and

Nuts have been shown to lower cholesterol by 1% for every 10 g (~10 almonds) eaten per day.

other nut products that only contain nuts and do not use added oils for processing (e.g. look for dry roasted almonds or almond butters that list only "almonds" under Ingredients), as the extra oil will contain unnecessary calories and may not be of the same quality. If nuts are not able to be consumed, substitute seeds instead.

Viscous Fiber

Eggplant

Eggplant is high in viscous fiber. It also contains many different vitamins and minerals such as:

- Vitamin C
- B6
- Potassium
- Magnesium

Two types of eggplant are commonly available: large black eggplant used primarily in Italian cooking and long, slender light purple eggplant commonly used in Japanese cuisine and often referred to as "Japanese eggplant." Japanese eggplants tend to be less bitter. To reduce bitterness in Italian eggplants, try cooking them thoroughly until very soft, as this action caramelizes the sugars making it taste sweeter.

A well cooked eggplant is less bitter.

Berries

Berries are a source of viscous fiber. Strawberries were added as a later addition to the Portfolio diet to increase the antioxidant content and palatability of the diet. Feel free to experiment with other kinds of forest berries including raspberries, blueberries and blackberries. These berries are good sources of:

- Vitamin C
- Fiber
- Potassium
- Folate

Okra

Okra is also an excellent source of viscous fiber and packed with nutrients such as:

- Vitamin A
- Vitamin C
- Calcium
- Magnesium
- Potassium
- Iron

When in season, okra can be found fresh in the produce section, and year-round in the frozen section. If you cannot find okra in your local supermarket, try Asian specialty food stores. Okra produces a slimy substance when cooked owing to its high viscous fiber content. This texture is unpleasant to some people but can be minimized with different cooking methods. If boiling okra, try adding a tsp. of white vinegar to the water and boiling the okra whole. Frying okra also helps keep the slimy texture to a minimum. For both frying and boiling preparation methods do not overcook the okra. Just like pasta, okra is best served "al dente" or slightly firm.

Oat Bran

Oat bran is an excellent source of the viscous fiber beta-glucan, which is associated with cholesterol lowering benefits. Oat bran consists of the outer coating of the grain where the beta-glucan is found, making it a concentrated source. In order to produce flaked oats or quick cooking oats, the bran is partially or entirely removed. Flaked oats still contain beta-glucan although very little of the bran usually remains making them a much poorer source. Compared to flaked oats, oat bran contains slightly more protein and is more nutrient rich. Oat bran is high in:

- Potassium
- Phosphorus
- Folate
- Selenium
- Omega-6 fatty acids

Oat bran can be a bit chewier than flaked oats and may require more soaking/cooking time depending on your preference.

Barley

Barley is high in antioxidants, low in fat, and has one of the lowest glycemic indexes of all the grains. It also contains beta-glucan, the same viscous fiber found in oats. Simply put, barley is a little known superfood. It is rich in nutrients such as:

Pearl barley is a great rice substitute.

- Zinc
- Manganese
- Selenium
- Calcium
- Potassium
- Magnesium
- Folate
- B6
- Iron

Barley can be substituted for rice or other grains wherever they are used. It is commonly available in three forms: hulled, pot, and pearl barley.

- Hulled barley has a rich and chewy texture and contains the most fiber out of all of the three forms. However, it takes the longest to cook and so is best used when more time for meal preparation is available.
- Pot barley is most commonly used in soups to add heft and thickness to the dish.
- Pearl barley is the most common type of barley available and works well as a rice substitute.

Barley can easily be made in large batches and frozen for later use.

Psyllium

If using powdered psyllium, use 1/2 of what the recipes in this book require.

Psyllium is made from the seeds of a plant called *Plantago ovata,* native to India. It is an extremely concentrated source of viscous fiber. Pysllium can be purchased on its own or as an additive in fortified foods. For example, pysllium is one of the main ingredients in Metamucil®, used to treat constipation. Be aware that psyllium comes in powdered and husk forms. If using powdered psyllium, use ½ of what the recipes in this book require. Owing to its high viscous fiber content, psyllium will thicken any mixture to which it is added. As such, it's a great egg substitute in baking products and when added in small amounts to bread mixes, helps to leaven the bread.

Plant-Based Protein

Legumes

Legumes like chickpeas, lentils, kidney beans, and pinto beans are all low glycemic index sources of protein and fiber. This means they are suitable for individuals who have or are at risk for diabetes. They are high in diverse range of nutrients including:

Rinse canned beans to get rid of compounds that cause gas.

- Iron
- Folate
- Manganese
- Potassium
- B6

Legumes are a great source of plant-based protein and are also very economical. You can find legumes dried, or canned. All of these options are equally good sources of legumes, although dried beans may require some additional planning. For example, dried chickpeas and kidney beans need to be soaked overnight. On the other hand, dried lentils, black eyed peas, and split peas do not require soaking and take about 20 min to 1 h to prepare depending on the recipe. If you buy legumes in a can, make sure to rinse them thoroughly as the liquid in which they sit may cause gas and may contain high levels of salt.

Soybeans

Soybeans have the highest amount of protein of any vegetable. They are low in saturated fat and high in:

Look for soy products with 6–8 g of protein per serving.

- Iron
- Calcium
- Magnesium
- Potassium
- Phosphorus
- Folate
- Monounsaturated and polyunsaturated fatty acids

Soybeans can be found canned, dried, or frozen. Remember to rinse canned soybeans to avoid gas after consumption. Dried soybeans need to be soaked overnight, but after cooking can be frozen and eaten at a later time. To save time try cooking in large batches. Fresh soybeans are often frozen and sold as "edamame." To cook, they only need to be boiled or microwaved for a couple of minutes, making them an easy addition to soups or salads for a little extra protein. Edamame also makes a great snack on it's own and can be eaten plain with a pinch of salt or seasoning.

Soy Milk

Many different flavors of soy milk are available. Be aware that flavored soy milk will contain extra sugar and calories. Check the protein content of your soy milk before purchasing. Look for soy milk that contains 6–8 g of protein per serving.

Soy Yogurt

Soy yogurt is made from soy milk and typically contains active bacterial cultures similar to those found in regular yogurt. It can be purchased in health food stores or the health section of larger grocery stores. Check the protein content and look for brands with 7 g or more per serving.

Soy Meat Analogues

Many soy-based meat alternatives are available in the refrigerated sections of health food stores or in the health food section of grocery stores. These substitutes range from "chicken" tenders to ground "beef," soy burger, soy hot dogs, and soy deli slices. Look for soy products with 6–8 g of protein per serving.

Soy Mayonnaise

Soy mayonnaise is a great alternative to regular mayonnaise. Its made with either silken tofu or soy milk instead of egg yolk. As it does not contain any eggs and is entirely plant based, it does not contain any cholesterol. You can find soy mayonnaise in health food stores or in the health food section of larger grocery stores.

Textured Vegetable Protein

Textured vegetable protein (TVP) contains small soy nuggets that are made from soy flour. They can be used in foods such as soups, casseroles, or stews. TVP is a common ingredient in soy burgers, vegetarian chicken nuggets, and other meat analogs. They can be found in health food stores or the health food section of grocery stores.

Tofu

Tofu is made from soybeans and comes in many different forms and flavors. Extra firm tofu contains the most soy protein, the active component shown to lower cholesterol. Extra firm and firm tofu work well in dishes like soups or stir-fry's. Silken and soft tofu are best when a smooth and creamy texture are desired, like in smoothies or desserts. Medium or firm tofu can be used in dishes that require an egg-like texture, like tofu scrambles.

Plant Sterols

Plant sterols are naturally occurring compounds found in certain grains, fruits, and vegetables. They are added to foods such as margarine and juice to improve cardiovascular health benefits from consumption. A commonly found margarine fortified with plant sterols is Becel ProActiv®. If you cannot find plant sterol margarine good substitutes include supplements, or enriched juices. Vegan Becel™ can be used for baking, canola baking. Canola and high oleic sunflower for stove-top cooking, and olive oil for drizzling. These oils are all high in monounsatuturated fatty acids and contain some plant sterols. However, these substitutes will contain a lower amount of plant sterols than fortified foods. As a result, it will be harder to meet your daily goal for plant sterol intake. If possible, try to obtain fortified foods like margarine, juices or supplements.

TIPS FOR KEEPING ON TRACK WITH THE PORTFOLIO DIET

Fig. 4.1 outlines some general rules to help you stay on track when following the Portfolio diet. As mentioned above, starting a new diet can be a challenge and starting a diet filled with unfamiliar foods even more so. The tips in Fig. 4.1 are meant to give you a sense of how to plan your meals throughout the day

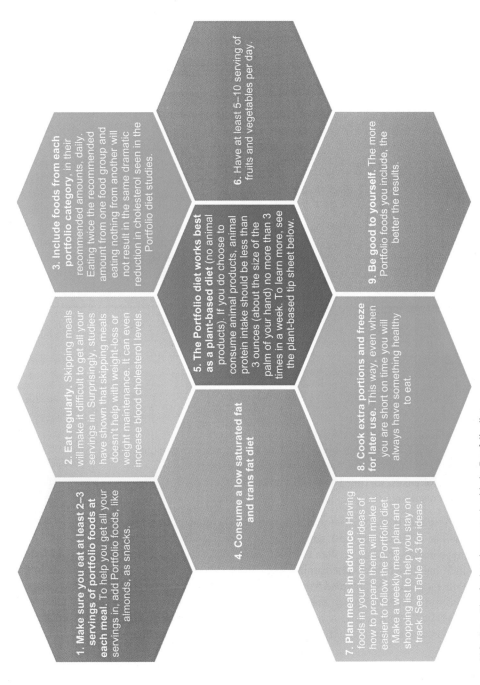

FIG. 4.1 Tips for keeping on track with the Portfolio diet.

so you can reach your goal for each of the Portfolio food groups. They involve everyday activities as well as ideas for how to plan ahead. Along the same lines, included at the end of the chapter is a 3-day eating guide (Table 4.3). The guide is there to give you a sense of how you can structure your meals and snacks. It can also be helpful to follow when you initially start the diet. Remember, although it may be complicated at first, you will be planning meals in no time.

TIPS FOR EATING A PLANT-BASED DIET

A plant-based diet is one that contains no animal products. Animal products include eggs, meat such as fish, chicken, beef, and pork as well as dairy products like cheese, milk, and butter. When starting a plant-based diet there are two main concerns individuals often have. The first surrounds how to replace the previously mentioned animal products. The second involves how to avoid deficiency, specifically for protein, B12 and iron.

As discussed above, dairy and meat products can easily be substituted for soy-based alternatives. Eggs can likewise be substituted.

Eggs. There are many commercially available egg substitutes which can be found as powders in the baking aisle or refrigerated in the egg section of most grocery stores. For baking, store bought varieties can be found in the health sections of many grocery stores, however you can also easily make your own. Egg substitutes have the added bonus of minimizing cholesterol intake.

Tip

Leftovers from dinner make a quick and easy lunch the following day.

---

1 EGG=
- 1 tbsp. psyllium, chia, or flax mixed with 3 tbsp. water.
- 1/3–1/4 cup of mashed banana, applesauce, or silken tofu, experiment with different amounts and find the one that works for you.
- Particularly for cakes, mixing 1 tsp. apple cider vinegar with 2 tsp. baking powder will result in a fluffy texture that is usually found in eggs.

Lower dietary cholesterol intake with egg-substitutes.

Protein: Protein deficiency is very rare in developed countries and eating a varied diet will be sufficient for most people to get the protein they need. The phrase "A complete source of protein" is often thrown around in conversation about plant-based foods. Proteins are made up of building blocks called amino acids. Some of these amino acids can be made in the human body while others need to be obtained from food. Humans have nine essential amino acids that need to be obtained from food. Meat and other animal products contain all nine essential amino acids making them what is known as a "complete source of protein." However, complete proteins sources are also present in some plant foods, like soy beans[2] or quinoa. It is also important to note that it is not necessary for a single

---

[2] While there has previously been some discussion over the safety of the soy bean, it is now generally agreed that it is a safe and healthy food. For more information on the topic see Box 3.3 in Chapter 3.

food to contain a complete source of protein, as foods can be combined to create complete sources of protein. For example, legumes and grains have complementary amino acids. When combined in the same meal, or even eaten in the same day, they provide all 9 essential amino acids and form a complete source of protein.

Plant-based foods high in protein include:

- Tofu (10 g/1/2 cup)
- Tempeh (15 g/1/2 cup)
- Lentil (8.8/1/2 cup)
- Peanuts (20 g/1/2 cup)
- Broccoli (16 g/1 bunch weighing about 606 g)

Iron: Iron is an important nutrient and plays the vital role in transporting oxygen from the lungs to other tissues of the body. Unlike foods from animal sources, plant-based foods do not naturally contain what is known as heme iron. Heme iron from hemoglobin in animal blood is absorbed into the body at a much greater rate than nonheme iron found in plant-based foods. This difference in absorption means that in order to achieve the same levels of iron in your bloodstream, you would need to eat more nonheme iron than heme iron. This is why individuals often worry if they will be able to consume enough iron on a plant-based diet. However, the high absorption rate of heme iron may not be such a good thing as iron has prooxidative effects and can result in tissue damage, inflammation as well as increased risk of CVD [1]. Eating a varied diet will be sufficient for most people to receive the iron intake they need. Plant-based foods high in iron include:

- Dark leafy greens, spinach and kale
- Legumes, soybeans and lentils
- Nuts/seeds, almond and pumpkin seeds

B12: When following a plant-based diet it would be useful to have your physician assess your B12 levels annually to ensure all is well. B12 supplements are readily available (made by microbes) and can be used to increase B12 levels if required.

While recently plant-based diets have gotten an increasing amount of attention in the west, this eating pattern is by no means a new concept. Plant-based diets have been traditionally consumed for over 500 years by groups such as Buddhists and Jains. Plant-based diets have numerous health benefits including the reduction of chronic disease like diabetes, cancer, and heart disease which are discussed further in Chapter 6. At first glance, a plant-based diet can seem limiting and complicated. However, through trial and experimentation plant-based diets can offer a new horizon of tastes through combinations of fruits, vegetables, seeds, nuts, soy products, cereals, whole grains, and legumes. See Fig. 4.2 for more tips on ensuring optimal nutrition while following a plant-based diet.

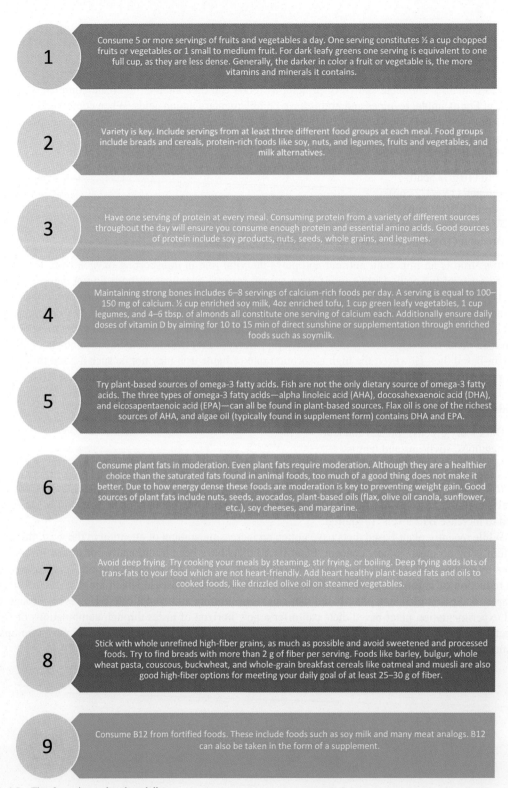

1 Consume 5 or more servings of fruits and vegetables a day. One serving constitutes ½ a cup chopped fruits or vegetables or 1 small to medium fruit. For dark leafy greens one serving is equivalent to one full cup, as they are less dense. Generally, the darker in color a fruit or vegetable is, the more vitamins and minerals it contains.

2 Variety is key. Include servings from at least three different food groups at each meal. Food groups include breads and cereals, protein-rich foods like soy, nuts, and legumes, fruits and vegetables, and milk alternatives.

3 Have one serving of protein at every meal. Consuming protein from a variety of different sources throughout the day will ensure you consume enough protein and essential amino acids. Good sources of protein include soy products, nuts, seeds, whole grains, and legumes.

4 Maintaining strong bones includes 6–8 servings of calcium-rich foods per day. A serving is equal to 100–150 mg of calcium. ½ cup enriched soy milk, 4oz enriched tofu, 1 cup green leafy vegetables, 1 cup legumes, and 4–6 tbsp. of almonds all constitute one serving of calcium each. Additionally ensure daily doses of vitamin D by aiming for 10 to 15 min of direct sunshine or supplementation through enriched foods such as soymilk.

5 Try plant-based sources of omega-3 fatty acids. Fish are not the only dietary source of omega-3 fatty acids. The three types of omega-3 fatty acids—alpha linoleic acid (AHA), docosahexaenoic acid (DHA), and eicosapentaenoic acid (EPA)—can all be found in plant-based sources. Flax oil is one of the richest sources of AHA, and algae oil (typically found in supplement form) contains DHA and EPA.

6 Consume plant fats in moderation. Even plant fats require moderation. Although they are a healthier choice than the saturated fats found in animal foods, too much of a good thing does not make it better. Due to how energy dense these foods are moderation is key to preventing weight gain. Good sources of plant fats include nuts, seeds, avocados, plant-based oils (flax, olive oil canola, sunflower, etc.), soy cheeses, and margarine.

7 Avoid deep frying. Try cooking your meals by steaming, stir frying, or boiling. Deep frying adds lots of trans-fats to your food which are not heart-friendly. Add heart healthy plant-based fats and oils to cooked foods, like drizzled olive oil on steamed vegetables.

8 Stick with whole unrefined high-fiber grains, as much as possible and avoid sweetened and processed foods. Try to find breads with more than 2 g of fiber per serving. Foods like barley, bulgur, whole wheat pasta, couscous, buckwheat, and whole-grain breakfast cereals like oatmeal and muesli are also good high-fiber options for meeting your daily goal of at least 25–30 g of fiber.

9 Consume B12 from fortified foods. These include foods such as soy milk and many meat analogs. B12 can also be taken in the form of a supplement.

FIG. 4.2 Tips for eating a plant-based diet.

TABLE 4.3   Three-Day Diet Plan: Provides ~2000 kcal, 40 g Soy Protein, 20 g Soluble Fiber, 20 g Nuts, and 2 g Plant Sterols

	Day 1	Time Saving Variations	
Breakfast	1 sv. Overnight Muesli (pg. xx) ½ cup orange juice with 1 tbsp. psyllium *Optional: tea or coffee with soy milk*	1 sv. Oat Bran (see Chapter 5) ½ cup orange juice with 1 tbsp. psyllium *Optional: tea or coffee with soy milk*	
Snack	3 cups whole of strawberries		
Lunch	"Egg" Salad (pg. xx) with arugula 2 sl. Oat Bran Psyllium Bread (pg. xx)		
Snack	1 cup soy milk ½ cup baby carrots		
Dinner	2 sv. Tofu Stir Fry (pg. xx) ½ cup ckd. barley mixed with ½ tsp. plant sterol margarine		
Snack/ dessert	1 sv. Apple Pie Mixed Nuts (pg. xx) 1 cup soy milk	1/2 cup mixed nuts 1 cup soy milk	
	Day 2	**Time Saving Variations**	
Breakfast	1 srv. Peanut Butter Smoothie (pg. xx) *Optional: tea or coffee with soy milk*	2 slc. Oat Bran Psyllium Bread (pg. xx) or whole wheat toast 1 1/2 tbsp. peanut butter (mixed with 1 tbsp. psyllium and 1 tsp. plant sterol margarine) 1 ½ cups soy milk *Optional: tea or coffee with soy milk*	1 sv. Leftover Overnight Muesli from Day 1. ½ cup soy milk *Optional: tea or coffee with soy milk*
Snack	1 Oat Bran Muffin (pg. xx) ½ tsp. plant sterol margarine 1/2 cup soy milk	Small handful of nuts Oat bran (pg. xx)	
Lunch	3 cups whole raspberries 1/2 sv. Miso Soup (pg. xx) served with 1 cup ckd. Konjac noodles	1 cup edamame 1 cup ckd. Konjac noodles 1 tsp. soy sauce	2 sv. Leftover Tofu Stir Fry (pg. xx)
Snack	¾ cup Edamame		
Dinner	2 sv. Ratatouille ½ cup ckd. barley mixed with ½ tsp. plant sterol margarine	Pressure ckd. eggplant and zucchini served with lemon and olive oil ½ cup barley with ½ tsp. plant sterol margarine	
Snack/ dessert	1 cup soy milk ½ cup raspberries 3 sv. Almond Cluster Cookies (pg. xx)	1 cup of soy milk Small handful of nuts	

Continued

TABLE 4.3 Three-Day Diet Plan: Provides ~2000 kcal, 40 g Soy Protein, 20 g Soluble Fiber, 20 g Nuts, and 2 g Plant Sterols—Cont'd

	Day 3 (Weekend)	Time Saving Variations	
Breakfast	2 Oat Bran Pancakes (pg. xx) 1 cup strawberries 2 tsp. plant sterol margarine 1/2 cup orange juice with 1 tbsp. psyllium *Optional: tea or coffee with soy milk*	1 sv. Strawberry Shortcake Smoothie (pg. xx) 2 sl. Oat Bran Psyllium Bread (pg. xx) 2 tsp. plant sterol margarine *Optional: tea or coffee with soy milk*	2 Leftover Oat Bran Muffins from Day 2 2 tsp. plant sterol margarine 1/2 cup orange juice with 1 tbsp. psyllium *Optional: tea or coffee with soy milk*
Snack	1 cup soy milk		
Lunch	1 sv. Creamy Broccoli Almond Soup 1 sl. Toasted Oat Bran Psyllium Bread 1 tsp. plant sterol margarine 2 sv Tabbouleh	½ cup of mixed nuts Hummus (pg. xx) on 2 sl. oat bran Psyllium Bread (pg. xx) (suggestion: add tomato and lettuce) 1 cup of soy milk	
Snack	1 sv. Baba Ganoush with fresh veggies for dipping 1 cup blackberries	1 cup orange juice with 1 tbsp. psyllium 1 Pear	
Dinner	1 sv. Okra Dahl (pg. xx) ½ cup ckd. barley		
Snack/ dessert	1 sv. Flan with berries 1 cup soy milk plus 2 tsp. psyllium	1 1/2 cups soy milk plus 2 tsp. psyllium Ginger Fuyu Persimmon (pg. xx)	

REFERENCE

[1] Eftekhari MH, Mozaffari-Khosravi H, Shidfar F, Zamani A. Relation between body iron status and cardiovascular risk factors in patients with cardiovascular disease. Int J Prevent Med 2013;4(8):911–6. Available from: http://www.ncbi.nlm.nih.gov/pubmed/24049617.

Chapter 5

Recipes

The following chapter consists of Portfolio diet recipes developed for this book and a selection of those used in the original studies. The original Portfolio diet recipes were developed by Caroline Brydson and team. These recipes used in the original studies were given out to patients to aid with the incorporation of Portfolio foods in the diet and in this way contributed to the results observed. In order to distinguish original Portfolio diet recipes a banner running across the top of each page identifies them. Some of the original recipes have been adapted to fit current recommendations and the style of this book. However, for the most part they remain intact.

Try incorporating dietary Portfolio foods, particularly those heavy in viscous fiber, slowly into the diet.

As mentioned in "Tips and Tricks,' in the beginning, try incorporating dietary Portfolio foods, particularly those heavy in viscous fiber, slowly into the diet. Introducing these foods gradually into your diet will give your body a chance to adjust and reduce any discomfort you may experience from the shift. This could mean, for example, starting off with the aim of getting a quarter of the recommended amount of viscous fiber. Foods like psyllium can easily be reduced in these recipes in order to accomplish this. Do not be afraid to experiment with the levels you eat, remembering that even small amounts of dietary Portfolio components can achieve positive health outcomes.

Incorporating these recipes into your diet will hopefully make the transition to the Dietary Portfolio an easy one and help to serve as inspiration in making it your own.

The Portfolio Diet for Cardiovascular Disease Risk Reduction. https://doi.org/10.1016/B978-0-12-810510-8.00005-4

BREAKFAST

SMOOTHIES

Tip

Try adding psyllium, oat bran, soymilk, and or nut butter to your favorite smoothies.

STRAWBERRY SHORTCAKE SMOOTHIE

Servings: 2
Time: 5 min

> 2 cups (300 g) frozen strawberries
> ½ cup (25 g) oat bran
> 2 large dates (Medjool preferred)
> 2 tsp vanilla extract
> 2 tbps psyllium
> 1 ¾ cups (425 ml) soy milk

In a large blender, combine strawberries, oat bran, dates, vanilla and psyllium, blend until smooth adding soy milk slowly until desired consistency is reached.

Nutrition Information: Strawberry Shortcake Smoothie *Per Serving: 300 kcal*			
Plant based protein	9.7 g	Total fat	4.3 g
Total carbohydrate	56.7 g	– Saturated	0.5 g
Dietary fiber	13.0 g	– Monounsaturated	1.2 g
– **Viscous fiber**	**6.5 g**	– Polyunsaturated	2.3 g

GREEN MANGO SMOOTHIE

Servings: 2
Time: 5 min

> 2 cups (375 g) frozen mango
> ½ avocado
> ¾ cup (40g) fresh spinach (about 2 handfuls of spinach)
> 2 large dates (Medjool preferred)
> 2 tbsp psyllium
> 1¾ cup (375 ml) soy milk

In a large blender, combine mango, avocado, spinach, dates and psyllium, blend until smooth adding soy milk slowly until desired consistency is reached.

Nutrition Information: Green Mango Smoothie			
Per Serving: 375 kcal			
Plant based protein	7.0 g	Total fat	10.6 g
Total carbohydrate	69.0 g	– Saturated	1.5 g
Dietary fiber	15.9 g	– Monounsaturated	5.7 g
– **Viscous fiber**	**6.0 g**	– Polyunsaturated	2.8 g

PEANUT BUTTER SMOOTHIE

Servings: 2
Time: 5 min

> *3 heaping tbsp (35 g) natural peanut butter or unsweetened nut butter*
> *1 banana, frozen or fresh*
> *½ tsp cinnamon*
> *2 tbsp psyllium*
> *1½ cups soy milk*

In a large blender, combine peanut butter, banana, cinnamon, psyllium, blend until smooth adding soy milk slowly until desired consistency is reached.

Nutrition Information: Peanut Butter Smoothie			
Per Serving: 262 kcal and 20g of nuts			
Plant based protein	10.2 g	Total fat	14.6 g
Total carbohydrate	24.0 g	– Saturated	1.9 g
Dietary fiber	6.0 g	– Monounsaturated	0.8 g
– **Viscous fiber**	**2.5 g**	– Polyunsaturated	1.6 g

OVERNIGHT OATS

Servings: 5
Time: 10 min, chill overnight

Tip

Save any leftovers and store in the fridge up to 3 days. Simply freshen up bowl with some soy milk and enjoy.

---

1 medium (80 g) apple, grated
1 cup (155 g) blueberries (frozen)
⅓ cup (35 g) slivered almond
⅓ cup (55 g) raisins
1 cup (45 g) oat bran
½ cup (50 g) rolled oats
1 tbsp brown sugar
2 tbsp of psyllium
3 ½ cup (850 ml) soy milk

1. In a large bowl, mix the apple, blueberries, almonds, raisins, oat bran, rolled oats, brown sugar, psyllium and soy milk together.
2. Cover and place in fridge to chill overnight.
3. Add additional soy milk until desired consistency is reached.

Nutrition Information: Overnight Oats			
Per Serving: 315 kcal and 6 g of nuts			
Plant based protein	12.5 g	Total fat	8.0 g
Total carbohydrate	51.3 g	– Saturated	0.9 g
Dietary fiber	10.0 g	– Monounsaturated	3.6 g
– Viscous fiber	**3.5 g**	– Polyunsaturated	3.1 g

OAT BRAN

Time: 5 min
Servings: 1

Tip

Try using cold soy milk for a refreshing and fast breakfast during the hot summer months.

---

1 ½ cups (375 ml) soy milk
¾ cup (35 g) oat bran
2 tsp psyllium

1. In a small saucepan, bring soy milk to a simmer, stirring regularly.
2. Mix in oat bran and psyllium, then remove from heat.

Nutrition Information: Oat Bran *Per Serving: 306 kcal*			
Plant based protein	22.6 g	Total fat	10.8 g
Total carbohydrate	56.6 g	– Saturated	1.7 g
Dietary fiber	15.6 g	– Monounsaturated	3.1 g
– **Viscous fiber**	**6.7 g**	– Polyunsaturated	5.6 g

TOPPING SUGGESTIONS

RASPBERRY MINT

¼ cup (35g) raspberries
2–3 sprigs mint
1 tbsp cacao (or dark chocolate chips)

ALMOND AND BROWN SUGAR

1 tbsp ground almonds, toasted
1 tbsp brown sugar
1 tbsp large flake oats (optional for texture)

HAZELNUT MAPLE

1 tbsp roughly chopped hazelnuts
1 tbsp maple syrup

ALMOND BUTTER AND JAM

1 tbsp almond butter
½ tbsp strawberry jam

CINNAMON PEAR

1 (180 g) pear, diced
1 tsp cinnamon
2 tbsp pumpkin seeds

BANANA ON TOAST

Servings: 1
Time: 5 min

  1 banana cut into rounds
  A pinch of cinnamon
  2 tbsp peanut butter
  1 slice toasted bread

Suggestion

Use oat bran bread with psyllium (page 123) instead of whole wheat bread toast.

Nutrition Information: Banana on Toast			
Per Serving: 362 kcal and 25 g of nuts			
Plant based protein	12.7 g	Total fat	16.8 g
Total carbohydrate	46.4 g	– Saturated	2.5 g
Dietary fiber	7.8 g	– Monounsaturated	8.2 g
– **Viscous fiber**	**2.1 g**	– Polyunsaturated	5.1 g

MEDITERRANEAN STYLE TOFU SCRAMBLE

Servings: 4
Time: 15 min

Tip

To increase plant based protein intake, try firm tofu. Just add an additional ½ cup of water or vegetable stock per block of firm tofu and cook down to desired consistency.

½ cup (100 g) thin strips of red onion (about half a medium onion)
½ tsp turmeric
2 tsp olive oil
1 block (454 g) medium firm tofu

2 tbsp olive brine or vegetable stock
½ cup (100 g) roughly chopped Kalamata olives
1 (140 g) bag of spinach
Salt and pepper to taste

1. In a large frying pan cook onion in turmeric and oil over medium heat until onions are translucent, about 5 minutes.
2. Using your hands, crumble tofu into the frying pan.
3. Add olive brine and cook until water is mostly evaporated, about 5 minutes.
4. Finally, mix in spinach and olives. Cook for 2 minutes or until spinach is wilted.
5. Add salt and pepper to taste.

Suggestion

Serve with oat bran bread with psyllium (page 123) toast topped with plant sterol margarine.

Nutrition Information: Mediterranean Style Tofu Scramble
Per Serving: 151 kcal

Plant based protein	10.5 g	Total fat	8.9 g
Total carbohydrate	8.7 g	– Saturated	1.1 g
Dietary fiber	3.1 g	– Monounsaturated	4.5 g
– **Viscous fiber**	**0.2 g**	– Polyunsaturated	3.2 g

LEMONY EDAMAME AVOCADO TOAST

Servings: 2
Time: 10 min

> *4–5 (45 g) button mushrooms, sliced*
> *1 tsp olive oil*
> *1 (200 g) avocado*
> *½ cup (60 g) edamame*
> *½ tsp lemons zest*
> *2 tsp lemons juice*
> *2 tsp psyllium*
> *Salt and pepper to taste*
> *2 slices whole wheat toast*

1. In a medium frying pan, lightly fry mushrooms in olive oil and set aside.
2. In a small bowl mash avocado and edamame with lemon zest and juice.
3. Mix in psyllium with the mashed avocado and add salt and pepper to taste.
4. Serve avocado mixture on oat bran bread with psyllium bread (page 123) or whole wheat toast, top with the fried mushrooms.

Tip

Great for topping with whatever vegetables or nuts you have in the kitchen.

Nutrition Information: Lemony Edamame Avocado Toast
Per Serving: 300 kcal

Plant based protein	9.0 g	Total fat	19.4 g
Total carbohydrate	27.8 g	– Saturated	2.7 g
Dietary fiber	12.9 g	– Monounsaturated	11.7 g
– Viscous fiber	**3.5 g**	– Polyunsaturated	2.5 g

WALNUT OAT BRAN MUFFINS

Servings: 10
Time: 20–30 min

Dry

> *½ cup (65 g) roughly chopped walnuts*
> *2 cups (185 g) oat bran*
> *1 cup (125 g) all-purpose flour*
> *½ cup (100 g) brown sugar*
> *2 tsp baking powder*
> *½ tsp baking soda*
> *½ cup (85 g) raisins*
> *3 tbsp psyllium*
> *1 pinch of salt*

Nutrition Information: Walnut Oat Bran Muffins
Per Serving: 300 kcal and 6 g of nuts

Plant based protein	5.0 g	Total fat	14.2 g
Total carbohydrate	40.3 g	– Saturated	1.8 g
Dietary fiber	4.0 g	– Monounsaturated	4.8 g
– **Viscous fiber**	**2.0 g**	– Polyunsaturated	6.2 g
		Plant Sterols	1.1 g

Wet

> *1 ½ cups (375 ml) unsweetened soy milk*
> *2 tsp apple cider vinegar*
> *½ cup (110 g) plant sterol margarine, melted*
> *1 tsp vanilla extract*

Topping

> *2 tbsp roughly chopped old fashioned oats (or oat bran)*
> *1 tbsp brown sugar*

Tip

If whole wheat flour is used add an extra ½ cup of soymilk.

---

1. Preheat the oven to 350°F (175°C). Lightly grease a muffin tin.
2. Starting first with the dry ingredients, toast walnuts in a medium sized pan over medium-to-low heat, stirring occasionally until slightly browned. Set aside to cool.
3. In a large bowl to mix oat bran, flour, brown sugar, baking powder, baking soda, raisins, psyllium and salt. Add the cooled walnuts.
4. To prepare the wet ingredients, use a small bowl to combine soy milk, vinegar, plant sterol margarine and vanilla, then add to the dry ingredients, mixing sparingly until well combined. The mixture should be scoop-able but not too wet (add more soy milk or flour until desired consistency is reached).
5. Distribute the mixture evenly between the cups in the muffin tin. Fill the cups just up to the brim.
6. For the topping, mix the oats and sugar together in a small bowl, and sprinkle on the muffins.
7. Place in the oven and bake for 15–20 minutes or until fork comes out clean.

FLUFFY OAT BRAN PANCAKES

Servings: 8
Time: 25 min

Tip

Poke with fork to check to see if the center is cooked (fork will come out clean). If not cooked through, cover with lid to speed up the process.

Wet ingredients
   2 tsp apple cider vinegar
   2 cups (500 ml) soy milk
   1 tbsp plant sterol margarine, melted

Dry ingredients
   ¾ cup (90 g) whole wheat flour
   ¾ cup (35 g) oat bran
   1 tbsp psyllium
   2 ½ tsp baking powder
   1 tsp brown sugar
   1–2 tsp vegetable oil or plant sterol margarine for cooking

> Variation:
> Add 1 cup of fresh or frozen blueberries to batter.

1. To prepare the wet ingredients, in a small bowl mix vinegar with soy milk, and plant sterol margarine and set aside.
2. To prepare the dry ingredients, use a larger bowl and combine flour, oat bran, psyllium, baking powder and sugar.
3. Add wet ingredients to dry ingredients and stir until combined. Do not over mix.
4. Let sit for 5 minutes or until the mixture has thickened.
5. Ladle mixture in generous dollops (about 1/4 cup) on to a hot, lightly oiled, frying pan.
6. Cook until bubbles start to form then flip (about 1–2 minutes on each side).

Nutrition Information: Fluffy Oat Bran Pancakes
Per Serving: 191 kcal

Plant based protein	6.9 g	Total fat	7.9 g
Total carbohydrate	28.1 g	– Saturated	0.8 g
Dietary fiber	5.5 g	– Monounsaturated	3.6 g
– Viscous fiber	**1.3 g**	– Polyunsaturated	2.9 g

Tip

The longer the batter sits the more moisture it will absorb, add up to ½ cup more soymilk to keep consistency the same. Batter should be thin enough to pour easily and not more than ¼ inch thick in the pan.

Suggestion

Warm frozen blueberries and add with liquid as sauce along with fresh fruit and plant-based whipped cream.

ALMOND BUTTER AND JELLY SANDWICH

It's a tasty, quick sandwich for grown-ups—with toasty almond butter, 100% fruit spread and heart-healthy oat bran bread. Try this as an on-the-go breakfast or anytime-of-day snack.

2 slices psyllium & oat bran bread
1 tbsp. almond butter
1 tbsp. 100% fruit spread

1. Spread slice of bread with almond butter and fruit spread. Assemble sandwich and serve.

Find more heart-healthy almond recipes at www.almondsarein.com.
Courtesy of the Almond Board of California

Nutrition Information			
Entire recipe contains:			
Calories	284	Fat	12 g
Protein	9 g	Polyunsaturated fat	3 g
Saturated fat	1 g	Fiber	3 g
Monounsaturated fat	7 g		
Cholesterol	0 mg		
Carbohydrates	38 g		

APPLE OAT BRAN MUFFINS

Makes the Portfolio diet more portable—enjoy as a treat or break-fast-on-the-go. Spread with plant sterol margarine these muffins contain a generous helping of all four *pillars* of the Portfolio diet.

2 cups (200 g) oat bran
⅔ cups (70 g) soy flour
4 tbsp. (20 g) psyllium husk
1 tbsp. baking powder
¼ cup (25 g) ground almonds
1 tbsp. ground cinnamon
1 tsp. salt
1 apple, grated or ½ cup unsweetened applesauce
½ cup brown sugar
2 cups (500 ml) soy beverage
1 tsp. vanilla extract
2 tbsp. (15 g) sliced almonds

1. Measure and blend together oat bran, soy flour, psyllium, baking powder, ground almonds, cinnamon, and salt.
2. Measure brown sugar, soy beverage, vanilla, and apple in a separate bowl.
3. Combine wet with dry ingredients and mix thoroughly.
4. Portion the batter into greased or paper lined muffin pans. Garnish with almonds.
5. Bake at 375°F for 20 to 25 min or until golden brown.

Tip

Psyllium and soy flour are available at bulk, health stores and grocery health food sections.

Use vegan Becel margarine if plant sterol margarine is not available.
Makes 12 muffins.
Recipe tested by Kathy Galbraith.

Nutrition Information
Each muffin (1/12 recipe) made with defatted soy flour contains:

Energy	163 kcal	Monounsaturates	2.2 g
Protein	8.1 g	Saturates	0.6 g
Soy protein	**4.5 g**	Total carbohydrate	27.8 g
Total fat	4.2 g	Dietary fiber	5.6 g
Polyunsaturates	1.5 g	**Viscous fiber**	**2.6 g**

BANANA PEACH SMOOTHIE

Put a tiger in your tank! Combine psyllium and soy milk with any fruit in season.

1 banana
1 peach, or other fruit
1 cup fortified soy beverage
1 tsp. psyllium husk
Add 2–3 dates for sweetness

1. Blend and enjoy.

Tip

Psyllium is available at health food stores and supermarkets.

Makes 1 smoothie.

Nutrition Information *Entire recipe contains:*			
Energy	230 kcal	Monounsaturates	1.1 g
Protein	11.2 g	Saturates	1.0 g
Soy protein	**9.3 g**	Total carbohydrate	63.9 g
Total fat	5.6 g	Dietary fiber	6.7 g
Polyunsaturates	3.0 g	**Viscous fiber**	**1.9 g**

AUNT NETTIE'S TOFU SCRAMBLE

This recipe originated from Aunt Nettie at Vegetarians in Paradise. The ingredients have been substituted to accommodate the Portfolio diet, but we kept the name.

½ red pepper, chopped
½ green pepper, chopped
1 small tomato, chopped
½ red onion, thinly sliced lengthwise
1 small red or white potato, shredded
2 tbsp extra virgin olive oil
1 package (350 g) extra firm tofu, crumbled
¼ cup (25 g) roasted sliced almonds*
1 tbsp Curry powder
Salt and pepper to taste

1. Heat a large skillet or wok over medium heat.
2. Add oil, peppers, onions, and potatoes and sauté for 3 to 4 min.
3. Add tomato, curry powder, and crumble tofu. Mix well and simmer for another 3 min.
4. Garnish with almonds over top.

Tip

Crumble tofu with your fingers. See www.vegparadise.com for more recipes.

* To roast almonds, place in a preheated 350F oven, stirring once or twice for approximately 10 minutes or until brown.

Makes 4 servings.

LUNCH

VEGGIE BLT

Servings: 1
Time: 10 min

Tip

Try adding avocado.

Tempeh and toppings

2oz (60 g) plain tempeh (about 5 slices)*
½ tbsp vegan mayonnaise and/or hummus (page 114)
1 tsp psyllium
2 slices toasted whole wheat or oat bran bread with psyllium (page 123)
1–2 leaves romaine lettuce
1–2 slices tomato

Tempeh sauce

2 tsp maple syrup
1 tbsp barbeque sauce (page 130)

1. In a frying pan on medium heat, cook the tempeh with maple syrup and barbeque sauce until browned on both sides, approximately for 5 minutes.
2. In a small bowl mix psyllium and vegan mayonnaise.
3. To whole wheat bread apply vegan mayonnaise and psyllium spread, layer tempeh, lettuce and tomato.

*Shortcut

Pre-flavored tempeh can also be used for even quicker sandwich making. Simply fry tempeh until crispy, omitting the tempeh sauce and construct sandwich to your liking.

Nutrition Information: Veggie BLT
Per Serving: 442 kcal

Plant based protein	21.5 g	Total fat	15.7 g
Total carbohydrate	53.0 g	– Saturated	2.3 g
Dietary fiber	6.9 g	– Monounsaturated	7.0 g
– **Viscous fiber**	**2.0 g**	– Polyunsaturated	4.6 g

SESAME TOFU LETTUCE WRAP

Servings: 1
Time: 10 min

> ¼ package (115 g) firm tofu, cubed into bite-sized pieces
> 2 tsp olive oil or canola oil
> ½ tsp pepper
> 1 tsp sesame oil
> 1 tbsp vegan mayonnaise
> 1 tsp psyllium

½ tbsp soy sauce
1 large romaine lettuce leaf
⅓ cup (70 g) diced tomato (about half a medium tomato)
1 tsp toasted sesame seeds

1. Preheat oven to 350°F (175°C).
2. On a parchment lined baking tray place tofu and coat it with oil and pepper. Lay tofu out so they are not touching each other. Bake for 20 minutes or until slightly brown, turning halfway.
3. In a small dish mix sesame oil, mayonnaise, psyllium, and soy sauce together.
4. Add tofu to lettuce leaf with tomato topping with sauce and sesame seeds.

Variation:
Instead of lettuce, try a whole wheat flour wrap. Simply chop up lettuce leaf and place inside the wrap with the rest of ingredients.

Nutrition Information: Sesame Tofu Lettuce Wrap
Per Serving: 300 kcal

Plant based protein	11.2 g	Total fat	24.3 g
Total carbohydrate	9.6 g	– Saturated	2.1 g
Dietary fiber	3.8 g	– Monounsaturated	11.7 g
– **Viscous fiber**	**1.0 g**	– Polyunsaturated	8.8 g

CHICKPEA "EGG" SALAD

Servings: 4
Time: 10 min

Tips

Baba gnoush can work as a great substitute for vegan mayonnaise.
Don't have chickpeas? Just double the tofu instead.

½ cup (100 g) chickpeas
½ package (230 g) firm tofu
¼ cup (60 g) vegan mayonnaise or mustard
1 tsp apple cider vinegar
1 tbsp olive oil
2 tsp mustard
2 tsp psyllium
2 tbsp lemon juice
½ tsp lemon zest
¼ cup (40 g) finely diced red onion (about half a small onion)
Salt and pepper to taste

1. In a large bowl mash chickpeas with fork.
2. Press tofu between two kitchen towels to get rid of excess water.
3. Crumble tofu into bowl with chickpeas using your hands.
4. Mix in mayonnaise, apple cider vinegar, olive oil, mustard, psyllium, lemon juice, lemon zest, and onion.
5. Season with salt and pepper.

Suggestion

Serve on toasted whole wheat or oat bran bread with psyllium (page 123) and top with arugula.

Nutrition Information: Chickpea "Egg" Salad *Per Serving: 200 kcal*			
Plant based protein	8.0 g	Total fat	16.4 g
Total carbohydrate	8.0 g	– Saturated	1.4 g
Dietary fiber	3.0 g	– Monounsaturated	9.9 g
– Viscous fiber	**1.8 g**	– Polyunsaturated	4.8 g

EDAMAME TABBOULEH

Servings: 4
Time: 30 min

Tip

Make sure to thoroughly dry parsley before chopping, this will allow for a finer cut.

1 cup (250 ml) boiling water
½ cup (110 g) bulgur/cracked wheat
1 ½ cup (40 g) finely chopped parsley, without stems
1 ½ cup (240 g) edamame beans
½ cup (110 g) finely diced onion
1 cup (200 g) finely diced tomato (about 1 large tomato)
½ cup (75 g) finely diced cucumber
2 tbsp lemon juice
1 tbsp olive oil
Salt to taste

Variation:
Try using barley
instead of bulgur
to increase viscous
fiber content.

1. In a small bowl combine 1 cup of boiling water with the bulgur and cover. Let sit for 20 minutes or until all of the water has been absorbed.
2. While the bulgur is cooking prepare the parsley, edamame, onion, tomato and cucumber.
3. After the bulgur has finished cooking, quickly cool the bulgur placing it in a fine sieve and running it under cold water. Finish by pressing the bulgur firmly against the sieve to remove any excess water.
4. In a large serving bowl combine the bulgur, parsley, edamame, onion, tomato, and cucumber.
5. Drizzle with lemon juice and olive oil. Add salt to taste.

Tip

Cover and keep in the fridge for up to 4 days for a quick lunch or snack.

Nutrition Information: Edamame Tabbouleh			
Per Serving: 170 kcal			
Plant based protein	8.6 g	Total fat	6.0 g
Total carbohydrate	23.4 g	– Saturated	0.6 g
Dietary fiber	6.1 g	– Monounsaturated	2.6 g
– Viscous fiber	**0.9 g**	– Polyunsaturated	0.5 g

CREAMY BROCCOLI ALMOND SOUP

Servings: 4
Time: 20–25 min

¾ cup (75 g) ground almonds
1 tsp plant sterol margarine
1 cup (150 g) finely diced onion (about 1 medium onion)
2 cups (500 ml) vegetable stock
1 cup (250 ml) water
1 ½ cups (250 g) chopped broccoli florets (about 1 head of broccoli)
½ package (170 g) cubed medium firm tofu
1 tbsp nutritional yeast flakes
Salt and pepper to taste

> Caution: Be careful when blending hot liquids. If blender is not equipped to handle higher temperatures let cool before blending.

1. In a large soup pot on medium-to-low heat toast ground almonds, stirring regularly until slightly browned, approximately 2–3 minutes. Put in a small bowl and set aside.
2. To the soup pot, add plant sterol margarine and onions, cooking until onions are translucent.
3. Add vegetable stock, water, broccoli, tofu and the toasted almonds to the soup pot. Cover and let simmer until broccoli is soft, approximately 10 minutes.
4. Transfer the soup to a blender, adding nutritional yeast flakes, salt, and pepper. Blend until smooth adding additional water until the desired consistency is reached.

Tip

Serve with whole wheat or oat bran bread with psyllium (page 123) topped with plant sterol margarine.

Nutrition Information: Creamy Broccoli Almond Soup
Per Serving: 171 kcal and 18 g of nuts

Plant based protein	10.4 g	Total fat	12.3 g
Total carbohydrate	9.1 g	– Saturated	1.1 g
Dietary fiber	6.0 g	– Monounsaturated	6.7 g
– **Viscous fiber**	**0.5 g**	– Polyunsaturated	3.8 g
		Plant Sterols	0.1 g

MAC AND "CHEESE"

Servings: 6
Time: 45–50 min

Tip

Try adding some steamed broccoli to the prepared mac and "cheese."

Pasta

1 (550 g) package whole wheat macaroni noodles

"Cheese" Sauce

⅔ cup (100 g) finely diced onion (about half a mid-sized onion)
2 ½ tbsp plant sterol margarine divided
2 cups (500 ml) water
1 (227 g) block of medium firm tofu
¾ cup (100 g) peeled and chopped carrots into ½ inch (1 cm) round pieces (about 1½ large carrots)
⅓ cup (20 g) nutritional yeast flakes
2 tsp psyllium
½ tsp salt or to taste

Almond topping

½ cup (50 g) ground almonds
½ tbsp nutritional yeast flakes crumbled
1 tbsp psyllium
Pinch of salt or to taste

Nutrition Information: Mac and "Cheese"
Per Serving: 444 kcal and 8 g of nuts

Plant based protein	22.3 g	Total fat		11.6 g
Total carbohydrate	72.1 g	– Saturated		1.4 g
Dietary fiber	12.2 g	– Monounsaturated		4.9 g
– **Viscous fiber**	**1.1 g**	– Polyunsaturated		3.9 g
		Plant Sterols		0.5 g

1. Cook the pasta according to the instructions on the box.
2. While the pasta is cooking prepare the cheese sauce. In a medium-sized pot sauté the onions in ½ tablespoon of plant sterol margarine, cooking until the onions are translucent.
3. Add two cups of water, the tofu, and the carrots to the pot with the onions. Cover and simmer for approximately 20 minutes or until the carrots are soft.
4. In a blender combine carrots, tofu and any remaining water from the pot along with the nutritional yeast, two tablespoons plant sterol margarine, pysyllium and salt. Blend until smooth, adding more water if necessary to reach a consistency similar to that of melted nacho cheese.
5. To make the topping, in a small pan on medium-to-low heat, toast the ground almonds stirring regularly, for approximately 2–5 minutes or until ground almonds become light brown. In a small bowl combine the toasted almond with the nutritional yeast, psyllium and salt.
6. Combine "Cheese" sauce with the pasta.
7. Sprinkle almond topping on pasta.

SPICY GLASS NOODLE SALAD

Serves: 6
Time: 15 min

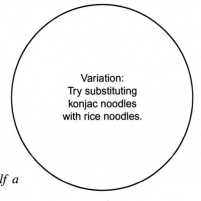

Variation:
Try substituting
konjac noodles
with rice noodles.

1 package (375 g) konjac noodles, rinsed and drained
3 cups (400 g) chopped red cabbage cut lengthwise into stripes
(about half a small cabbage)
1 package (460 g) firm tofu, cubed into bite-sized blocks
2 tsp grated fresh ginger
1 tbsp soy sauce
2 tsp sesame oil
2 tsp sambal oelek or other chili paste (optional)
1 ½ cups (150 g) fine ribbons of sliced cucumber (about half a
cucumber)

1. Prepare konjac noodles according to directions on package. Then rinse under cold water until they are cool to the touch.
2. In a large bowl mix cabbage, tofu, ginger soy sauce, sesame oil and sambal oelek. Add the noodles and let marinate in fridge while you prepare the cucumber.
3. Take cucumber and cut into fine ribbons. This can be done using a spiralizer or a potato peeler, peeling the cucumber lengthwise to get long strands.
4. Add cucumber to the bowl with cabbage and noodles and mix.

Nutrition Information: Spicy Glass Noodle Salad *Per Serving: 90 kcal*			
Calories	95	Total fat	4.9 g
Plant based protein	7.2 g	– Saturated	0.6 g
Total carbohydrate	6.5 g	– Monounsaturated	1.5 g
Dietary fiber	3.6 g	– Polyunsaturated	2.7 g
– **Viscous fiber**	2.7 g		

POT O' RISOTTO

This rice-free Risotto pot o' barley is a hearty classic.

> *1 tbsp. olive oil*
> *1 onion, finely chopped 1 clove garlic, chopped*
> *1 tsp. dried thyme*
> *1 cup (200 g) barley, dry*
> *4 cups vegetable stock*
> *1 cup (180 g) frozen soybeans (edamame)*
> *¾ cup (120 g) frozen, or fresh cooked okra, sliced*

1. Heat up a large pot over medium high heat. Add the olive oil, onion, garlic, and thyme. Sauté the onion until transparent.
2. Add the barley and stir. Begin adding the stock, 1 cup at a time, stirring until the barley has absorbed most of the liquid.
3. Cook the barley over medium-low heat until the barley is tender—30 minutes or more. Cover and stir often.
4. Add the soybeans and okra, and more water if necessary. Cook for another 5 min.

Tip

Look for frozen soybeans in health food stores and large super-markets. We used 1–2 bouillion vegetable cubes and water as stock.

Variation: to make Pot O' Barley soup, double the number of cups of vegetable stock.
Makes 4 servings.

Nutrition Information
Each 400 g serving (1/4 recipe) contains:

Energy	265 kcal	Monounsaturates	3.6 g
Protein	12.0 g	Saturates	1.5 g
Soy protein	**5.2 g**	Total carbohydrate	50.1 g
Total fat	7.8 g	Dietary fiber	13.6 g
Polyunsaturates	2.0 g	**Viscous fiber**	**4.3 g**

FRIED BARLEY

Forget fried rice, try using leftover barley instead.

2 cups (400 g) cooked barley, well drained and dried
3 cloves garlic, chopped
½ inch piece ginger root, finely chopped
2 green onions, chopped
5 sprigs of cilantro, stem finely chopped, leaves intact
⅓ package (115 g) extra firm tofu
½ cup frozen peas
2 tbsp Canola
2 tsp Sesame oil
2 tbsp Light soy sauce, or Tamari

1. Break tofu into small pieces. Marinate soy sauce.
2. Heat a wok over high heat and add the canola oil. Sauté the garlic, ginger, green onion, and cilantro stems. Quickly add barley and continue to stir-fry for a few min.
3. Add tofu and frozen peas. Sprinkle the sesame oil and soy sauce over the barley.
4. Garnish with cilantro leaves.

   Makes 4 servings.

Nutrition Information
Each 175 g serving (1/4 recipe) contains:

Energy	206 kcal	Monounsaturates	5.0 g
Protein	8.6 g	Saturates	1.0 g
Soy protein	**4.8 g**	Total carbohydrate	25.0 g
Total fat	10.2 g	Dietary fiber	5.5 g
Polyunsaturates	3.5 g	**Viscous fiber**	**2.0 g**

SOY-CHILI

Chili that's perfect for the Portfolio diet. Substitute soy-beans for kidney beans and ground soy for beef. Try it.

1 tbsp extra virgin olive oil
2 cups (230 g) soy ground round or 2 crumbled soy burgers
5 cloves garlic, chopped
1 onion, chopped
1 stalk celery, sliced
19 oz can soybeans, or other bean, drained and rinsed
28 oz can tomatoes, diced
2–4 tbsp. chili powder, mild
½ cup water or vegetable broth
Salt and pepper to taste

1. Heat the oil in a pot over medium high heat. Add the onion, garlic and celery, and sauté for 1–2 min.
2. Add the chili powder, ground soy, tomatoes, and soybeans. Add broth.
3. Bring chili ingredients to a boil, then reduce heat and simmer covered for about 30 min.
4. Adjust taste with salt and pepper.

Makes 4 servings.
Recipe tested by Le Commensal Restaurant: www.commensal.ca

Nutrition Information
Each 400 g serving (1/4 recipe) contains:

Energy	332 kcal	Monounsaturates	4.5 g
Protein	30.0 g	Saturates	1.7 g
Soy protein	**23.6 g**	Total carbohydrate	39.2 g
Total fat	13.4 g	Dietary fiber	16.4 g
Polyunsaturates	5.0 g		

BLACK BEAN BARLEY SALAD

A simple yet colorful dish that stands out in any meal or buffet.

2 cups (400 g) cooked barley, cooled
½ each red and green pepper, diced
19 oz can black beans, drained and rinsed well
1 tbsp. extra virgin olive oil
1–½ lemons, juiced
Fresh basil to taste
Salt and pepper to taste

1. Sauté peppers in 1 tbsp. oil for 1 min.
2. Combine all ingredients together in a large bowl.
3. Mix well and adjust seasonings to your taste.

   Variation: Add 2 diced tomatoes, 1 small diced avocado, ¼ tsp. hot pepper flakes and 2 sliced green onions.
   Makes 4 servings.
   Receipe variations tested by Kathy Galbraith, RD

Nutrition Information
Each 280 g serving (1/4 recipe) contains:

Energy	287 kcal	Saturates	1.7 g
Protein	11.0 g	Total carbohydrate	44.8 g
Total fat	11.8 g	Dietary fiber	10.6 g
Polyunsaturates	2.5 g	**Viscous fiber**	**2.0 g**
Monounsaturates	7.0 g		

SNACKS/ SMALL PLATES

SOY BEAN DIP

Servings: 4
Time: 5 min

> *1 cup (120 g) edamame beans*
> *½ (80 g) small cucumber*
> *⅓ cup (8 g) mint leaves (about a handful)*
> *¼ cup (65 ml) soy milk or ¼ block (115 g) of soft tofu*
> *1 tbsp lemon juice*
> *1 tbsp olive oil*
> *2 tbsp water*
> *Salt to taste*

Tip

If edamame is frozen, make sure to thaw in microwave before blending.

In a blender or a food processor, combine edamame, cucumber, mint, soy milk, lemon juice, olive oil, water and salt and mix until smooth. Serve with your favorite bread, cracker, or veggies.

Suggestion

To increase soluble fiber try mixing in 2 teaspoons of psyllium or serve with oat bran bread with psyllium (page 123).

Nutrition Information: Soy Bean Dip *Per Serving: 75 kcal*			
Plant based protein	3.3 g	Total fat	5.2 g
Total carbohydrate	2.8 g	– Saturated	0.5 g
Dietary fiber	1.7 g	– Monounsaturated	2.8 g
– **Viscous fiber**	**0.8 g**	– Polyunsaturated	0.5 g

BABA GANOUSH

Servings: 4
Time: 1 h

4–5 peeled cloves of garlic (or around one small head of garlic)
2 (1200 g) eggplants
⅓ cup (30 g) ground almonds
1 tbsp almonds butter (or ¼ cup (35 g) of whole almonds, smoothly ground)
1 tbsp lemon juice
1 tsp psyllium
1 tsp olive oil
Salt to taste

1. Preheat the oven to 400°F (205°C).
2. Using tin foil, tightly wrap each eggplant.
3. Drizzle garlic cloves with olive oil and wrap with tin foil.
4. Take a baking sheet and place the wrapped garlic and eggplants on it, bake in the oven for 30–45 minutes. or until both eggplant and garlic are very soft. A fork should easily slide through the skin when the eggplant is ready and garlic should be easily mashed.
5. While the eggplant and garlic are cooking, use a large frying pan to toast the ground almonds over medium-to-low heat, stirring occasionally until brown (2–5 minutes).
6. Remove eggplants from oven, cut into rounds, removing and discarding the stem.
7. Using a blender, blend eggplant, garlic, ground almonds, almond butter, lemon juice, psyllium, olive oil and salt. Blend until smooth.

Suggestion

Serve with fresh veggies oat bran bread with psyllium (page 123) or try using as a creamy sauce on barley.

Nutrition Information: Baba Ganoush *Per Serving: 150 kcal and 10 g of nuts*			
Plant based protein	5.2 g	Total fat	7.5 g
Total carbohydrate	20.4 g	– Saturated	0.8 g
Dietary fiber	9.5 g	– Monounsaturated	4.6 g
– **Viscous fiber**	**2.6 g**	– Polyunsaturated	1.7 g

CURRIED ALMOND HUMMUS

Servings: 5 servings
Time: 5–10 min

1 can (540 ml) chickpeas
1 tsp lemon juice or apple cider vinegar
1 ½ tsp curry powder
2 tsp olive oil
1 ½ tbsp almond butter (or ¼ cup (25 g) of whole almonds smoothly ground)
1 tbs roughly chopped onion
Salt to taste

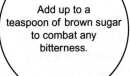

Add up to a teaspoon of brown sugar to combat any bitterness.

1. Drain chickpeas, making sure to set aside the liquid and, if so desired, a handful of chickpeas which can be used as garnish.
2. Using a blender, combine chickpeas, vinegar, curry powder, olive oil, almond butter and onion. While blending, add 2–3 tbsp chickpea liquid until mixture is smooth in texture.
3. Salt to taste.

Suggestion

Top with olive oil, whole chickpeas, paprika, and parsley.

Nutrition Information: Curried Almond Hummus			
Per Serving: 115 kcal and 3 g of nuts			
Plant based protein	5.4 g	Total fat	5.3 g
Total carbohydrate	13.0 g	– Saturated	0.6 g
Dietary fiber	4.4 g	– Monounsaturated	2.6 g
– **Viscous fiber**	**1.9 g**	– Polyunsaturated	1.6 g

EGGPLANT BITES

Servings: 4
Time: 35–40 min

Bites

   ½ cup (120 g) hummus
   2 tbsp mustard
   ¼ cup (25 g) ground almonds
   1 tbsp psyllium
   ¼ cup (32 g) nutritional yeast flakes
   ¼ cup (12 g) oat bran
   2 tsp dried oregano
   1 tsp dried thyme
   2 (850 g) Japanese eggplants, cut into rounds
   Pinch of salt

Topping/Dip

   2 tomato's diced finely
   ¼ small onion (75 g) diced finely
   Handful of fresh basil chopped into ribbons

> Variation: Italian eggplant can be substituted for Japanese eggplant. This size eggplant makes for less of a finger food and more of a side to a meal.

1. Preheat the oven to 350°F (175°C).
2. To make the bites, in a small bowl mix hummus and mustard together.
3. In a separate small bowl, mix ground almonds, psyllium nutritional yeast flakes, oat bran, oregano and thyme.
4. In a large, microwave safe bowl, microwave eggplant for 4 minutes.
5. Coat eggplant rounds in hummus/mustard mixture, then dip in oat bran/almond mixture so that they are fully coated.
6. Bake on parchment paper lined tray for 25–30 minutes or until the top of the eggplant is slightly browned.
7. For the Topping, use a small bowl and mix the tomato, onion and basil. Serve as an accompaniment to the eggplant bites.

Tip

Best served warm.

Nutrition Information: Eggplant Bites			
Per Serving: 210 kcal and 6 g of nuts			
Plant based protein	11.7 g	Total fat	6.6 g
Total carbohydrate	26.9 g	– Saturated	0.7 g
Dietary fiber	10.9 g	– Monounsaturated	3.2 g
– **Viscous fiber**	**3.7 g**	– Polyunsaturated	1.9 g

APPLE PIE MIXED NUTS

Servings: 6
Time: 2–4 h (7 min with shortcut)

Shortcut

Replace fresh apples with pre-dried apples cut into small pieces.

2 medium apples
2 tsp granulated sugar
¼ cup (40 g) toasted hazelnuts
¼ cup (35 g) toasted almonds
⅓ cup (42 g) toasted walnuts
½ tsp cinnamon
¼ cup (42 g) raisins

1. Preheat the oven on lowest temperature (150°F [65°C]).
2. Core and slice apples about ¼ inch (6 mm) thick.
3. Sprinkle granulated sugar on the top.
4. Place in slices on an oven tray lined with parchment paper and bake for 2–4 hours (depending on how crispy you like them) flipping halfway.
5. In a large frying pan over medium heat, roast the hazelnuts, almonds, and walnuts until light brown, mixing regularly for about 5 minutes.
6. Cut apples into bite-sized chunks and mix with nuts, cinnamon, and raisins.

Nutrition Information: Apple Pie Mixed Nuts
Per Serving: 150 kcal and 18 g nuts

Plant based protein	3.4 g	Total fat	10.7 g
Total carbohydrate	13.8 g	– Saturated	0.9 g
Dietary fiber	3.5 g	– Monounsaturated	5.0 g
– **Viscous fiber**	**1.1 g**	– Polyunsaturated	4.2 g

OAT ENERGY BAR

Servings: 10
Time: 25–30 min

> To prepare hazelnuts:
> bake for 10 min at
> 350°F (175°C)
> Let cool and
> then rub between
> hands to remove skins.

¾ cup (115 g) hazelnuts (roasted and peeled)
¼ cup (60 ml) fancy molasses
¾ cup (75 g) oats
1 cup (250 ml) unsweetened peanut butter

1. Preheat the oven to 350°F (175°C), and grease a 12×6 inch (30×15 cm) pan.
2. In blender or food processor, pulse hazelnuts until roughly chopped.
3. In a medium sized bowl, mix ground hazelnuts with the molasses, oats and peanut butter.
4. Put the batter in the greased pan.
5. Bake for 15 minutes.
6. Let it cool before cutting.
7. Store tightly wrapped at room temperature for up to 2 weeks. Or freeze for up to 1 month.

Tip

Bars become less sticky the longer they are air dried. To combat unwanted stickiness, let sit on the counter unwrapped for about 30 minutes to 1 hour.

Nutrition Information: Oat Energy Bar			
Per Serving: 255 kcal and 32 g of nuts			
Plant based protein	8 g	Total fat	19.0 g
Total carbohydrate	17.0 g	– Saturated	3.3 g
Dietary fiber	3.3 g	– Monounsaturated	4.8 g
– **Viscous fiber**	**1.2 g**	– Polyunsaturated	1.0 g

SALT AND VINEGAR EGGPLANT PICKLES

For the Pickle connoisseur, try this eggplant condiment. Combine with salad or a veggie side dish.

1 large eggplant (600 g)
1-½ cups white vinegar
1 cup water
½ tsp. oregano
1 tbsp. chopped parsley
1 clove garlic, chopped
1-2 tsp. olive oil
Salt to taste

1. Peel the eggplant and cut it into thick strips.
2. Boil water, vinegar, and salt. Add eggplant strips and cook until just tender.
3. Drain well. Add the remaining ingredients and chill.

   Makes 4 servings.

Nutrition Information			
Each 150 g serving (1/4 recipe) contains:			
Energy	72.3 kcal	Saturates	0.3 g
Protein	1.8 g	Total carbohydrate	14.3 g
Total fat	2.0 g	Dietary fiber	2.5 g
Polyunsaturates	0.3 g	**Viscous fiber**	**0.8 g**
Monounsaturates	1.5 g		

OAT BRAN BREAD WITH PSYLLIUM

Kathy Galbraith produced similar loaves for Dr. Jenkins cholesterol-lowering studies.

2– ¾ cups (280 g) oat bran
⅔ cup (90 g) gluten flour (wheat gluten)
¼ cup (20 g) psyllium husk
2 tbsp. sugar
1 tsp. salt
1 pkg (8 g) quick-rise yeast
2 cups warm water (120°F to 130°F)

> Variation:
> To make a crispier crust, try shaping dough into a ball and instead of pressing into a loaf tin, bake in cast iron (or any other oven safe pot) with lid. Cover for the first 30 minutes of baking and bake uncovered for the last 10–20 minutes.

1. Measure warm water, psyllium, salt and sugar into a mixing bowl. Add oat bran, gluten flour, dry yeast.
2. Lightly grease or spray a (4" × 8") loaf pan. Preheat oven to 350°F.
3. Begin to mix sloppy dough. It will become thick and dough-like. Sprinkle a small amount of oat bran on the table. Knead the dough to form a smooth loaf.
4. Shape dough into a pan. Cover and place in a warm spot for 10 to 15 min, until risen to top of bread pan.
5. Bake in a 350°F for 40–50 min. To brown crust, increase oven to 400°F for 5 min. Remove from pan to cool.

Makes 1 medium loaf. Recipe tested by Kathy Galbraith

This bread may be purchased at Kensington Natural Bakery, 460 Bloor Street West, Toronto (416) 534-1294.

Nutrition Information

Each 60 g slice (1/15 recipe) contains:

Energy	91 kcal	Saturates	0.3 g
Protein	7.6 g	Total carbohydrate	17.7 g
Total fat	1.2 g	Dietary fiber	5.5 g
Polyunsaturates	0.5 g	**Viscous fiber**	**3.5 g**
Monounsaturates	0.4 g		

BBQ BITES

Soy-chicken bites can be served anytime as a snack, or as a meat alternative on the grill.

2 soy-chicken burgers* (170 g) cubed
1 green pepper, diced large
½ eggplant (300 g) diced
10 (125 g) cherry tomatoes
1 cup Pineapple chunks (optional)
1 tsp. Canola
½ tsp. salt
½ tsp. garlic powder or soy sauce
Toothpicks or skewers
Montreal Chicken Spice, to taste

1. Dice eggplant and season with salt and leave for 30 min.
2. Rinse, drain and blot dry.
3. Coat fry pan with oil, sauté burger pieces, peppers, and eggplant on medium high heat for 2–4 min each side.
4. Sprinkle with spices, add tomato, pineapple and reduce heat. Arrange pieces on skewers and serve.

Tip

For the BBQ, combine seasonings with oil, coating each piece before skewering. Grill 2–4 min per side. Soy Sauce marinated tofu can be substituted for soy burgers.

Makes 2 servings.

Nutrition Information			
Each 340 g serving (1/2 recipe) contains:			
Energy	198 kcal	Monounsaturates	2.8 g
Protein	19.6 g	Saturates	0.7 g
Soy protein	**9.0 g**	Total carbohydrate	20.4 g
Total fat	6.0 g	Dietary fiber	4.0 g
Polyunsaturates	2.2 g	**Viscous fiber**	**0.8 g**

OAT BRAN COOKIES

Choose this cholesterollowering cookie chockful of chewy crunches.

½ cup Plant sterol margarine
1 cup sugar
1 cup (200 g) silken tofu (soft)
1 tsp. vanilla extract
2-¾ cups (280 g) oat bran
½ cup (60 g) barley flour
½ cup (50 g) ground almonds
2 tsp. baking powder
½ tsp. salt
Berry jam

1. Preheat oven to 375°F. Blend together margarine and sugar. Beat well. Add tofu and vanilla and mix until smooth.
2. Add the oat bran, barley flour, ground almonds, salt, and baking powder to the wet ingredients. Stir well to combine thoroughly.
3. Prepare a baking tray with cooking spray, and portion a tbsp of dough to form cookies. Using a wet spoon, make an indent on the top of each cookie. Fill the indent with a scant ½ tsp of fruit jam.
4. Bake cookies for 10 min or until edges begin to brown.

Makes 40 small cookies.
Recipe tested by Kathy Galbraith RD

Nutrition Information			
Each cookie (1/40 recipe) contains:			
Energy	71 kcal	Monounsaturates	1.4 g
Protein	1.8 g	Saturates	0.5 g
Soy protein	**0.3 g**	Total carbohydrate	9.6 g
Total fat	3.4 g	Dietary fiber	1.5 g
Polyunsaturates	1.3 g	**Viscous fiber**	**0.7 g**

DINNER

RATATOUILLE

Servings: 5
Time: 1 h

4 ½ cups (1000 g) chopped eggplant (about 2 eggplants)
1 tbsp olive oil
2 cups (300 g) chopped onion (about 2 mid-sized onions)
1 ½ tsp dried rosemary
1 tsp dried thyme
1 tsp dried oregano
2 cups (240 g) chopped zucchini (about 1 large zucchini)
½ cup (90 g) chopped red or orange pepper (about 1 pepper)
2 cups (400 g) diced tomato (about 4 mid-sized tomatoes)
1 can (800 g) crushed tomatoes with liquid
Salt and pepper to taste

> Variation:
> 3 cups of your
> favourite tomato
> sauce can be
> substituted for
> canned tomatoes.

Tip

To reduce the cooking time microwave eggplant with water for 1–2 minutes before beginning.

1. In a large pot over a medium heat, add olive oil and eggplant. Fry for approximately 5 minutes stirring regularly to make sure eggplant does not stick to the pan. Remove from pan and set aside.
2. To the pan now add the onions, rosemary, thyme, and oregano, and cook for 5 minutes, or until the onions are translucent.
3. Next, add back the eggplant to the pan along with the zucchini and pepper. Cook for approximately 5 more minutes.
4. Finally, add the fresh tomatoes and can of crushed tomatoes. Cover, reduce heat, and simmer for 45 minutes to 1 hour or until the eggplant is very soft, stirring occasionally.

Suggestion

Serve over barley or a large chunk of oat bran bread with psyllium (page 123).

Nutrition Information: Ratatouille
Per Serving: 161 kcal

Plant based protein	5.5 g	Total fat	4.1 g
Total carbohydrate	29.2 g	– Saturated	0.6 g
Dietary fiber	9.7 g	– Monounsaturated	2.2 g
– **Viscous fiber**	**3.1 g**	– Polyunsaturated	0.6 g

OVEN BAKED "MEATBALLS" AND ALMOND "PARMESAN"

Servings: 4
Time: 30 min

Use premade barbeque sauce in the "meatballs" to save time.

BBQ sauce

½ cup (125 g) finely diced onion (about 1 small onion)
2 tsp plant sterol margarine
1 tsp smoked paprika
1 tsp molasses
3 tbsp tomato paste
⅔ cup (150 ml) water

"Meatballs"

1 package (170 g) vegan ground round, or TVP
1 tsp caraway seeds
2 tbsp psyllium
½ cup (60 g) roughly chopped almonds
2 tbsp nutritional yeast flakes
2 tbsp chickpea flour

Almond "Parmesan"

⅓ cup (40 g) ground almonds
2 tbsp nutritional yeast flakes
1 tbsp psyllium
Pinch of salt or to taste

Pasta (optional)

½ package (200 g) whole wheat spaghetti
½ jar (200ml) tomato pasta sauce of your choice
1 bunch fresh basil

Nutrition Information: Oven Baked "Meatballs" and Almond "Parmesan" *Per Serving: 205 kcal and 14g of nuts*			
Plant based protein	16.1 g	Total fat	9.5 g
Total carbohydrate	17.9 g	– Saturated	0.8 g
Dietary fiber	9.3 g	– Monounsaturated	5.5 g
– **Viscous fiber**	**2.5 g**	– Polyunsaturated	2.3 g
		Plant Sterols	0.2g

1. Preheat the oven to 350°F (175°C).
2. If you are making the optional pasta, follow instructions on package to prepare.
3. To make the BBQ sauce, first fry the onion in 2 tsp of plant sterol margarine in a small pan over medium heat until onions are translucent and soft, approximately 5 minutes.
4. To the small pan add the smoked paprika, molasses, tomato paste, and water. Mix until well combined and heat until sauce is thick, approximately 5 minutes.
5. To make the "Meatballs", combine the vegan ground round, caraway seeds, psyllium, almonds, nutritional yeast and chickpea flour in a medium-sized bowl.
6. Add ¾ of the BBQ sauce to meatball mixture, setting ¼ aside to coat the outside of the formed meatballs.
7. Roll meatballs to the size of ping pong balls, then coat in the remaining BBQ sauce.
8. Bake on parchment paper lined baking tray. Position so that meatballs are not touching. Bake for 10–15 minutes turning halfway until meatballs are slightly browned on each side.
9. To make the almond Parmesan, toast ground almonds in a large pan on medium-to-low heat, stirring regularly until browned (2–5 minutes).
10. In a small bowl, combine the remaining Parmesan ingredients of nutritional yeast, psyllium, and salt with the toasted almonds.
11. Combine pasta and sauce, top with "meatballs", almond "Parmesan" and fresh basil.

CASHEW STIR FRY WITH CRISPY TOFU

Servings: 3
Time: 20–25 min

Tip

If chili paste used has a lot of added salt reduce soy sauce by half.

1 package (460 g) firm tofu cut in bite-sized chunks
2 tsp olive or canola oil, divided
¼ cup (60 ml) hoisin sauce
⅓ cup (80 ml) water
1 tbsp grated fresh ginger
1 tbsp soy sauce
2 tsp sambal oelek or chilli flakes (optional)
10–15 (170 g) asparagus spears with bottoms cut off (about 1–2 inches from base)
1 cup (150 g) roughly chopped red pepper
½ cup (42 g) sliced shallot cut into fine strips (about 1 shallot)
1 cup (75 g) sliced shitake mushrooms or cremini or portabello
⅔ cup (100 g) cashews
2 tsp sesame oil

Variation: Hoisin can be substituted by combining 1 tbsp rice vinegar (or light vinegar), 2 tbsp soy sauce, 1 tbsp water and 1 tbsp brown sugar.

1. Preheat the oven to 350°F (175°C).
2. In a large non-stick pan fry tofu in 1 teaspoon oil over medium heat, turning tofu so that sides are slightly browned, approximately 5 minutes. Remove from pan and set aside.
3. In a small bowl mix the hoisin, water, ginger, soy sauce and sambal.
4. In the large frying pan add fry the asparagus and red peppers in the remaining teaspoon of oil for 5 minutes.
5. Add scallions and fry for 2–3 minutes or until slightly translucent.
6. Add the mushrooms and cook until desired consistency is reached (approximately 5 minutes for al dente vegetables).
7. While vegetables are cooking, evenly space the cashews on baking tray and toast in oven for about 5 minutes or until cashews are slightly brown.
8. Mix vegetables, tofu and sauce in the small bowl together, drizzle with sesame oil and top with roasted cashews.

Suggestion

Serve over barley.

Nutrition Information: Cashew Stir Fry With Crispy Tofu
Per Serving: 325 kcal and 10 g of nuts

Plant based protein	18.5 g	Total fat	18.0 g
Total carbohydrate	24.2 g	– Saturated	2.6 g
Dietary fiber	4.6 g	– Monounsaturated	8.0 g
– **Viscous fiber**	**0.5 g**	– Polyunsaturated	6.6 g

OKRA DAL

Serving: 8
Time: 20–25 min

2 cups (250 g) finely diced onion (about 1 large onion)
1 ½ tbsp grated fresh ginger
1 tbsp plant sterol margarine or olive oil, divided
1 tbsp curry powder
2 tsp turmeric
2 tsp cumin seeds
2 tsp cinnamon
2 cups (450 g) finely diced tomatoes (about 3 large tomatoes)
1 tbsp sambal oelek or favorite chili sauce
1 ½ cups (300 g) dried red lentils
4 cups (1 L) water
½ package (230 g) firm tofu, crumbled
1 cup (250 ml) soy milk
1 tbsp psyllium
Salt to taste
3 ½ cups (375 g) frozen okra
Small bunch of cilantro (optional)

Variation:
For a thinner
sauce, try
cubing tofu
instead
of crumbling it.

1. In a large sauce pan over medium heat sauté onion and ginger in ½ tablespoon plant sterol margarine until onions are translucent, approximately 5 minutes.
2. Next add the curry powder, turmeric, cumin seeds and cinnamon and cook for another 2–3 minutes.
3. Add the tomato and sambal oelek and cook for another 2–3 minutes.
4. Next, add lentils, water, tofu, soy milk, psyllium and salt to taste and cook for another 15–20 minutes or until lentils are soft.
4. While the Dal is simmering, in a separate pan, fry the okra in the remaining ½ tablespoon of plant sterol margarine. Cook for 5–10 minutes stirring occasionally so that the okra is slightly browned.
5. Combine okra with the Dal and top with cilantro if desired.

Suggestion

Serve over barley.

Nutrition Information: Okra Dal *Per Serving: 245 kcal*			
Plant based protein	16.1 g	Total fat	4.8 g
Total carbohydrate	33.1 g	– Saturated	0.6 g
Dietary fiber	8.8 g	– Monounsaturated	1.6 g
– Viscous fiber	**3.5 g**	– Polyunsaturated	2.4 g

NUT LOAF

Servings: 6
Time: 1 h

Tip

Do not be scared away by the number of pans or lengthy list of ingredients, this is a versatile recipe great for using up any vegetables nearing the end of their shelf life.

Pan 1: Lentils

2 cups (500 ml) vegetable stock
½ cup (100 g) dried brown lentils
1 bay leaf

Pan 2: Vegetables

1 cup (200 g) finely diced onion (about 1 mid-sized onion)
1 cup (100 g) finely diced mushrooms
1 cup (110 g) grated carrot
1 cup (30 g) fresh spinach (about 3 large handfuls)
⅓ cup (60 g) black olives chopped roughly
⅓ cup (20 g) roughly chopped sun dried tomatoes
2 tbsp tomato paste
1 cup (125 g) roughly chopped, toasted walnuts
2 tbsp lemon juice
1 cup (50 g) whole meal bread crumbs
1 cup (135 g) ground almond
2 tbsp nutritional yeast flakes
Olive oil for cooking

Pan 3: Red wine gravy

½ cup (75 g) finely diced onion (about 1 small onion)
1 tsp plant sterol margarine
½ tsp sage (dried)
1 tsp thyme (dried)
1 tsp rosemary (dried)
2 tsp flour
½ cup (125 ml) red wine
1 cup (250 ml) vegetable stock

Pan 1: Lentils

1. In a medium pan bring vegetable stock to a boil.
2. Add lentils and bay leaf to the pan. Cook until the water has mostly boiled off, about 20 minutes. Remove bay leaf and drain excess water.

Pan 2: Vegetables

3. Preheat the oven to 375°F (190°C).
4. While lentils are cooking, in a large pot, sauté onion on medium high heat until translucent, for approximatively 5 minutes.
5. Next, add mushrooms and carrot, cooking for another 5 minutes.
6. Add spinach, reduce temperature, and cook until wilted and thoroughly mixed.
7. Add olives, sun dried tomatoes, tomato paste, walnuts, lemon juice, bread crumbs, ground almond and nutritional yeast flakes, keeping some ground almond and breadcrumbs aside for topping.
8. Add cooked lentils from pan 1: Lentils into the pan with your vegetables, pan 2: Vegetables and mix together.
9. Transfer mixture into a 1 L (2-pint) loaf pan, lined with wax paper, pressing it firmly in.
10. Top with remaining ground almond, and bread crumbs and cover with tin foil. Bake for 30 minutes.
11. Remove tin foil and bake for a further 15 minutes uncovered.

Pan 3: Red Wine Gravy

12. While loaf is baking, prepare the sauce.
13. In a small pan over medium-high heat, sauté onions with sage, thyme, and rosemary in plant sterol margarine, until onions are translucent, approximately 5 minutes.
14. Add flour and mix until well combined.
15. Add red wine and vegetable stock and simmer on medium high heat, stirring occasionally for 10–15 minutes or until sauce is thick.
16. Serve as gravy over the nut loaf or on the side.

Nutrition Information: Nut Loaf
Per Serving: 414 kcal and 30 g of nuts

Plant based protein	14.7 g	Total fat	20.9 g
Total carbohydrate	39.5 g	– Saturated	2.3 g
Dietary fiber	9.2 g	– Monounsaturated	8.3 g
– **Viscous fiber**	**1.4 g**	– Polyunsaturated	10.1 g

SWEET POTATO PIE

Servings: 6
Time: 1 h 45 min

Crust

>*½ cup (150 g) whole wheat flour*
>*½ tbsp nutritional yeast flakes*
>*Pinch of salt*
>*3 tbsp plant sterol margarine*
>*2 tbsp very cold water*

Filling

>*1 cup (200 g) finely diced onion (about 1 medium onion)*
>*1 tbsp curry powder*
>*2 ½ cups (500 g) peeled and diced sweet potato (about 2–3 small sweet potatoes)*
>*¾ cup (240 g) drained and rinsed chickpeas (about ½ can)*
>*1 cup (250 ml) vegetable stock*
>*1 cup (250 ml) water*
>*2 tbsp nutritional yeast flakes*
>*Salt to taste*
>*¾ cup (125 g) peeled and diced sour apple (about, 1 medium Granny Smith apple)*
>*⅓ cup (65 g) raisins*
>*2 tbsp lemon juice*
>*1 cup (125 g) roughly chopped walnuts*
>*1 ½ tbsp psyllium*
>*Olive or canola oil for cooking*

Topping

>*¼ cup (35 g) slivered almonds*

Tip

Keep plant sterol margarine as cold as possible. This will create a flakier crust.

Crust

1. In a large bowl mix flour, nutritional yeast flakes, and salt together.
2. Cut in plant sterol margarine by using a pastry cutter or two knifes slicing the mixture until it consists of small rounds, the size of a gravel.
3. Stir in water.
4. Form into a ball, cover with cling wrap, and chill in fridge (at least 45 minutes).

Tip

Cover surface with cling wrap before rolling out dough. This will make it easier to move once it is rolled out.

Filling

5. While dough is chilling, sauté onions over medium high heat with curry powder until translucent, approximately 5 minutes.
6. Add sweet potatoes, chickpeas, vegetable stock, and water. Cover with lid and cook for around 45 minutes or until potatoes are soft, adding more water if necessary.
7. Once the potatoes are soft, mash the mixture in the pot itself or place in a blender. Blend or mash until a chunky consistency is reached. Stir in nutritional yeast flakes, and salt to taste.
8. Add apple, raisins, and lemon juice.
9. Preheat the oven to 400°F (205°C).
10. Using a rolling pin, on a lightly floured surface, roll out the ball of dough (about a 20th of an inch [0.5 cm] thickness or less).
11. Grease a 10 inch (25 cm) diameter round pan, transfer dough to pan and trim the edges. You may have some left over dough depending on how thick you have made your crust. This dough can be saved in the fridge or freezer for a later date.
12. Using a fork, poke several holes in the crust and sprinkle walnuts evenly across the bottom of the pie shell. This position will toast the walnuts as well as prevent the crust from bubbling. Bake for 15 minutes.
13. Collect the now toasted walnuts from the base and combine with the rest of the filling. Add the psyllium.
14. Pack filling into the crust base and sprinkle with the topping of slivered almonds.
15. Bake for 20 minutes or until the nut topping is slightly browned.

Tip

Dough can be kept for up 4 days in the fridge and can be frozen for up to 6 months.

Nutrition Information: Sweet Potato Pie
Per Serving: 375 kcal and 24g of nuts

Plant based protein	10.5 g	Total fat	21.4 g
Total carbohydrate	40.0 g	– Saturated	2.2 g
Dietary fiber	9.9 g	– Monounsaturated	6.3 g
– **Viscous fiber**	**2.8 g**	– Polyunsaturated	11.5 g
		Plant Sterols	0.6 g

MISO SOUP

Serves: 4
Time: 15 min

6 cups (1.5 l) water
3 tbsp miso paste
2 tsp soy sauce
1 block (460 g) firm tofu cut into bite-sized cubes
2 cups (72 g) chopped Swiss chard leaves (around 5 leaves)
2 tbsp (10g) diced spring onion (around 2 sprigs)

Variation:
Try substituting
bok choy for Swiss
chard and adding a
handful of dried
seaweed.

1. In a large pot bring water to a boil on high heat.
2. Reduce burner to medium heat, add miso paste, soy sauce, tofu, and Swiss chard.
3. Cook for 5–10 minutes or until chard is soft.
4. Top each bowl with a sprinkle of with spring onion.

Nutrition Information: Miso Soup			
Per Serving: 132 kcal			
Plant based protein	12.3 g	Total fat	5.6 g
Total carbohydrate	9.0 g	– Saturated	0.7 g
Dietary fiber	2.5 g	– Monounsaturated	1.6 g
– Viscous fiber	**0.4 g**	– Polyunsaturated	3.3 g

SWEET POTATO BARLEY

Serves: 4
Time: 30 min

1 cup (200 g) barley
2 cup (500 ml) water
1 cup (150 g) peeled and cut into cubes sweet potato (about 1 medium sized sweet potato)
¾ cup (200 g) diced onion (about 1 medium-sized onion)
2 tsp olive oil or as needed
1 tsp dried ginger
1 cup (35 g) cut into ribbons Swiss chard (about 3 leaves)
2 tbsp ketjap manis
½ cup (15 g) mint with stems removed and cut into ribbons (about 1 large handful)

1. In a small pot combine barley with water and cook for 25 minutes or until barley is soft (it should still be chewy but not crunchy).
2. While barley is cooking, microwave the sweet potato in a mid-sized dish with about 1 inch (2.5 cm) of water at the bottom. Microwave for about 8 minutes or until soft, depending on your microwave. Mix halfway.
3. In a large frying pan, stir fry onion over medium-high heat with olive oil and ginger, until onions are translucent, approximately for 2–5 minutes.
4. Drain any excess water from sweet potatoes and add them to the pan with the onions. Fry the potatoes, turning occasionally, until sides are slightly browned, about 10 minutes. Add more oil for cooking as needed.
5. Add chard, ginger, ketjap, and barley to the frying pan and cook for 2–5 minutes more.
6. Top with mint.

Variation: Swiss chard can be substituted for other leafy greens such as kale or spinach.

Variation: Instead of ketjap manis use 1 tbsp soy sauce with 2 tsp fancy molasses and 1 tsp dried ginger.

Tip

Can be refrigerated and served as a cold salad.

Nutrition Information: Sweet Potato Barley
Per Serving: 257 kcal

Plant based protein	7.0 g	Total fat	3.1 g
Total carbohydrate	50.3 g	– Saturated	0.5 g
Dietary fiber	10.1 g	– Monounsaturated	1.9 g
– **Viscous fiber**	**3.1 g**	– Polyunsaturated	0.5 g

CREAMY MUSHROOM, BARLEY RISOTTO

Serving: 4
Time: 30 min

1 cup (200 g) finely diced onion (about 1 medium-sized onion)
1 cup (100 g) finely diced celery (about 2 stalks of celery)
5 sprigs of fresh thyme with stems removed (or two tsp dried thyme)
2 tsp olive oil
2 cups (125 g) diced cremini mushrooms (about 6 large mushrooms)
1 ½ cups (300 g) pearled barley
2 tbsp nutritional yeast flakes
3 cups (750 ml) vegetable stock
½ cup (125 ml) water
1 cup (75 g) chopped oyster mushrooms cut into rounds (about 5 mushrooms)
1 tsp plant sterol margarine
1 cup (25 g) finely chopped Italian parsley
2 tsp lemon zest (1 small lemon)
Salt and pepper to taste

Variation:
If oyster mushrooms are not available try Portobello or simply double the cremini.

Variation:
Try substituting 1/2 cup of dry white wine instead of vegetable stock to add some kick to your dish.

1. In a large pan sauté onions and celery with thyme in olive oil until onions are translucent (around 5 minutes).
2. Add cremini mushrooms and cook for another 2–3 minutes.
3. Next, mix in barley and nutritional yeast.
4. Once combined, add stock one cup at a time, waiting for most of the moisture to be absorbed before adding the next cup. Stir frequently (around 10–15 minutes).
5. At the same time, cook the oyster mushrooms in separate medium sized pan in plant sterol margarine, spreading out evenly in the pan. Cook until the mushrooms are slightly browned on each side. Set aside.
6. Once barley has finished cooking add parsley and lemon zest. There should still be some liquid visible in the bottom of the pan. Add up to ½ cup (125 ml) water to hydrate if this is not present.
7. Turn off heat, cover and let sit for 2–3 minutes.
8. Add oyster mushrooms on top.

Tip

When cooking mushrooms wait for a couple of minutes before stirring to let the outside sear, this will prevent them from getting too watery.

Nutrition Information: Creamy Mushroom Barley Risotto
Per Serving: 375 kcal

Plant based protein	15.6 g	Total fat	4.2 g
– **Soy protein**	**0 g**	– Saturated	0.7 g
Total carbohydrate	74.3 g	– Monounsaturated	2.0 g
Dietary fiber	17.4 g	– Polyunsaturated	0.9 g
– **Viscous fiber**	**4.5 g**	Plant Sterols	0.1 g

BARLEY, MUSHROOM CASSEROLE

Serve as a main meal or side dish.

> 2 tbsp. olive oil
> 1 large onion, chopped
> 1 stalk celery, diced
> ¼ cup green pepper, diced
> 1 cup (200 g) dry barley
> 1–½ cups sliced fresh mushrooms
> Salt and pepper to taste
> 2 cups vegetable stock
> ½ cup (70 g) roasted, roughly chopped almonds*

1. Sauté onion, celery, and green pepper in oil until soft.
2. Add barley and lightly toast.
3. Combine remainder of ingredients, aside from almonds, with onions, celery and pepper and pour into a casserole dish (greased with non-stick cooking spray).
4. Cover and bake at 350°F for 45 minutes.
5. Uncover and sprinkle almonds over top. Bake for another 15 minutes or until all liquid is absorbed.

Tip

Barley is available at bulk, health stores and grocery health food sections.

---

*To roast almonds, place them in a preheated 350°F oven for around 10 minutes, stirring once or twice. Makes 4 servings.

Nutrition Information
Each 300 g serving (1/4 recipe) contains:

Energy	306 kcal	Saturates	1.7 g
Protein	9.1 g	Total carbohydrate	46.6 g
Total fat	14.0 g	Dietary fiber	10.7 g
Polyunsaturates	2.6 g	**Viscous fiber**	**3.9 g**
Monounsaturates	9.2 g		

TENDER GINGER TOFU

An exquisitely simple and addictive dish developed by a favorite vegetarian buffet restaurant of Dr. Jenkins—Le Commensal Restaurant.

1 package (350 g) extra firm tofu
2 tbsp. (30 g) fresh ginger root, peeled, and grated
2 tbsp. Canola
2 tbsp. soy sauce or Tamari
¼ cup water
1 tsp. brown sugar

1. Cut tofu into ½ inch slices, about 10 slices. Place inside a shallow baking pan.
2. Mix together remaining ingredients and pour over tofu slices. Allow to marinate overnight or at least 1 h, turning once or twice.
3. Bake in oven at 350°F for 20–30 min or until the liquid has evaporated.

Makes 3 servings.
Recipe tested by Le Commensal Restaurant: www.commensal.ca.
In Toronto: 1 location, Elm and Bay St.
In Montreal: 8 locations; see back cover of cookbook for addresses

Nutrition Information			
Each serving (1/3 recipe) contains:			
Energy	219 kcal	Monounsaturates	6.0 g
Protein	19.0 g	Saturates	1.3 g
Soy protein	**19.0 g**	Total carbohydrate	7.6 g
Total fat	12.2 g	Dietary fiber	1.3 g
Polyunsaturates	4.6 g		

CHILI-NON-CARNE

Spoon over soy spaghetti or barley.

1 large eggplant, (600 g) diced
2 cups (230 g) soy ground round, or crumbled soy burger
19 oz can red kidney beans, drained and rinsed well
1 medium onion, chopped
1 clove garlic, chopped
2 tbsp. Canola
1 tbsp. fresh parsley, chopped
1 tbsp. dry fresh basil, chopped
4 cups fresh mushrooms, chopped
1 tbsp. soy sauce
1 tbsp. chili powder
½ tsp. salt
1 tsp. psyllium (husk)

1. Preheat oil in pan, add the chopped onions, garlic. Cook until tender.
2. Add the soy sauce, mushrooms and diced eggplant.
3. Add the spices, psyllium, parsley, basil and water as needed.
4. Once the eggplant is soft, stir in the ground soy, and the beans, and continue to cook for 5 min on medium heat.

Tip

Psyllium available in health food stores.

Makes 4 servings.
Recipe tested by Shannon Dixon

Nutrition Information			
Each 400 g serving (1/4 recipe) contains:			
Energy	276 kcal	Monounsaturates	4.4 g
Protein	22.4 g	Saturates	0.7 g
Soy protein	**9.1 g**	Total carbohydrate	42.4 g
Total fat	8.4 g	Dietary fiber	14.7 g
Polyunsaturates	2.6 g	**Viscous fiber**	**1.2 g**

NUTTY LENTIL LOAF

Chock full of protein, a delicious and satisfying substitute for meatloaf. Team with a green salad and a large glass of soy milk for a complete meal.

19 oz canned lentils, drained and rinsed
1 cup (100 g) almonds
1 medium onion, chopped
1 cup (100 g) oat bran
1 tsp. each: sage, celery seed, thyme, parsley
1 tbsp. soy sauce or Tamari
⅓ cup (75 g) olive oil
2 tbsp. psyllium husk mixed with 6 tbsp water
¾ cup warm vegetable stock

1. Preheat oven to 350°F. Combine warm stock and oat bran together in a large mixing bowl.
2. Pureé 1 cup of the lentils with the olive oil.
3. Add the remaining ingredients to the bowl. Mix well with hands or wooden spoon.
4. Scrape mixture into a greased or non-stick loaf pan. Bake for 1–½h.

Tip

Psyllium is available at bulk, health stores and grocery health food sections.

Makes 8 servings.
Recipe tested by Shannon Dixon

Nutrition Information
Each 130 g slice made with psyllium (1/8 recipe) contains:

Energy	265 kcal	Saturates	2.0 g
Protein	11.7 g	Total carbohydrate	26.8 g
Total fat	16.1 g	Dietary fiber	8.4 g
Polyunsaturates	2.5 g	**Viscous fiber**	**1.5 g**
Monounsaturates	10.9 g		

DESSERT

GINGER FUYU PERSIMMON

Servings: 2
Time: 25–35 min

Tip

Fuyu persimmons are a variety of persimmon that are not as sweet as other types of commonly available persimmon, such as Hachiya. In this recipe we use Fuyu persimmons, as their natural tarteness compliments the sweetness of the ginger syrup. You can recognize Fuyu persimmons as, unlike other varieties, they remain hard when ripe.

Ginger syrup

¾ cup (125 g) brown sugar
1 cup (250 ml) water
¼ cup (30 g) peeled and chopped fresh ginger

Almond Topping

1 tbsp roughly chopped almonds
1 persimmon, sliced in half and lightly scored in a grid pattern on cut side

Variation:
Try substituting
baked peaches
for persimmons.

1. In a small pot combine ingredients for ginger syrup (brown sugar, water and ginger). Simmer on medium heat for 20–30 minutes or until liquid thickens.
2. While the syrup is reducing, prepare the almond topping. In a small frying pan toast the almonds over medium-to-low heat, stirring occasionally until slightly brown.
3. Remove any large pieces of ginger from the syrup.
4. Spoon the ginger syrup onto each half of the persimmon.
5. Sprinkle with almonds.

Where to Find

Persimmons can be found in the fruit section of many fruit markets or in Chinese or Japanese grocery stores.

Nutrition Information: Ginger Fuyu Persimmon
Per Serving: 140 kcal and 5 g of nuts

Plant based protein	1.1 g	Total fat	2.4 g
Total carbohydrate	30.3 g	– Saturated	0.2 g
Dietary fiber	3.6 g	– Monounsaturated	1.4 g
– **Viscous fiber**	**1.4 g**	– Polyunsaturated	0.6 g

FLAN

Serving: 8
Time: 1 h 30 min

Tip

To remove flan from ramekins simply slide a knife around the edges and flip upside-down onto a plate.

Caramel

> *½ cup (100 g) brown sugar*
> *3 tbsp water*

Flan

> *½ cup (100 g) white sugar*
> *1 tbsp plant sterol margarine (melted)*
> *2 packages (600 g) soft tofu*
> *2 ½ tbsp cornstarch*
> *1 tbsp psyllium*
> *2 tsp vanilla extract*
> *1 ½ cups (375 ml) sweetened soymilk*
> *1 cup (100 g) blackberries or berries of choice (garnish)*

Experiment with different brands of tofu. Different brands may yield more or less batter and have different baking times.

1. Preheat the oven to 350°F (175°C).
2. To make the caramel sauce, mix together the sugar and water in a small pot, heat over medium heat. Allow to cook for 8 minutes without stirring. When sauce becomes sticky it is ready. Test by dipping a spoon in to the caramel make sure it has thickened. After being dipped the spoon should come out coated in caramel. Have 8 ramekins ready (each should hold around ⅓ (80 ml) cup liquid).
3. Evenly divide caramel into ramekins.
4. To make the flan, combine the white sugar, plant sterol margarine, tofu, cornstarch, psyllium, vanilla and soymilk in a blender and blend until smooth.
5. Fill ramekins with flan batter leaving around ½ an inch (1.5 cm) at the top.
6. Fill a 13-inch (33 cm) baking pan with two cups (500 ml) of water and place ramekins inside. Be careful not to get water in the ramekins.
7. Bake for 40 minutes or until the flan starts to slightly rise.
8. Let sit for at least 15 minutes to serve warm or chill for at least 2 hour or/overnight to serve cool.
9. Top with blackberries if desired.

Nutrition Information: Flan *Per Serving: 180 kcal*			
Plant based protein	4.3 g	Total fat	3.5 g
Total carbohydrate	34.4 g	– Saturated	0.3 g
Dietary fiber	2.3 g	– Monounsaturated	1.9 g
– Viscous fiber	**0.8 g**	– Polyunsaturated	1.0 g

PSYLLY BUNS

Servings: 9
Time: 2h 15min

Tip

Make sure that mixture is not hot when yeast is added so as not to kill it.

Dough

3 tbsp plant sterol margarine
1 ½ cup (350 ml) unsweetened soy milk
1 package (2 ¼ tsp) instant yeast
1 cup (80 g) psyllium
1 tbsp brown sugar
Pinch of turmeric for color (optional)
Pinch of salt
2 ½ cups (300 g) unbleached all-purpose flour

Filling

¾ cup (95 g) roughly chopped walnuts
1 cup (175 g) finely chopped dates
⅓ cup (80 ml) hot water
1 tsp vanilla extract
1½ tsp cinnamon

"Cream cheese" Frosting

½ cup (100 g) dairy free cream cheese (e.g. Daiya)
½ cup (60 g) icing sugar
1 tbsp of soy milk

Variation:
As an alternative to "Cream Cheese" Frosting, try making simple icing by mixing together 1/2 cup (60 g) icing sugar, 1/4 tsp vanilla and 1-1 1/2 tbsp of water in a small bowl.

Dough

1. To make the dough, first melt 3 tablespoons of plant sterol margarine in large microwave safe bowl, microwaving for about 20 seconds.
2. Add the soymilk and microwave for an additional 20 seconds until liquid is lukewarm.
3. Add the yeast, stirring with a fork until well combined. Let stand for 10 minutes to activate.
4. Next add the psyllium, brown sugar, turmeric and salt.
5. Finally add the flour. Briefly knead with your hands to ensure the mixture is combined and then form into a ball. Dough should be slightly sticky but easy to work with. Add more flour if needed.
6. Place dough ball into a large clean bowl. Cover with a towel and leave in a warm place for 1 hour or until the dough has doubled in size. Meanwhile make the filling.

Filling

7. To make the filling, first toast the walnuts in a large pan over medium-to-high heat, stirring occasionally until slightly brown.
8. In a blender combine dates, water, vanilla extract and cinnamon and blend until smooth, adding more water as needed to make a paste.
9. Stir in the walnuts to the paste.

Filling the buns

10. Roll out the dough on lightly floured surface into an 8.5 x 11 inch (20 x 30cm) rectangle, which is roughly the size of a sheet of paper. Dough should be thin (about a 20th of an inch [0.5 cm]).
11. Spread filling evenly onto the dough.
12. Starting at the long edge, roll dough into a tube.
13. Cut the roll into nine even segments.
14. Place rolls on a well-greased pan, with space between them so rolls do not touch each other.
15. Let the dough to rest for around 30 minutes in a warm place.
16. When dough is nearly finished resting, preheat the oven to 350°F (175°C).
17. Bake rolls for 25–30 minutes or until slightly browned on the top.
18. While the rolls are baking make the "Cream Cheese" frosting. Combine "cream cheese" alternative, icing sugar and soymilk together in a medium sized bowl and blend with hand mixer until smooth.
19. Apply "Cream Cheese" frosting liberally to warm buns.

Nutrition Information: Psylly Buns
Per Serving: 400 kcal and 8 g of nuts

Plant based protein	7.5 g	Total fat	17.0 g
Total carbohydrate	56.3 g	– Saturated	3.1 g
Dietary fiber	9.5 g	– Monounsaturated	3.5 g
– **Viscous fiber**	**5.5 g**	– Polyunsaturated	8.6 g
		Plant Sterols	0.4 g

FUDGY BROWNIES

Servings: 12
Time: 45 min

> ½ cup (115 g) plant sterol margarine
> 1 can (398 ml) white pinto beans (unsalted) drained and washed
> ¼ cup (60 ml) maple syrup (or 2 tbsp sugar dissolved in ¼ cup [60 ml] water)
> ¾ cup (130 g) semi-sweet (dark) chocolate chips
> ¾ cup (150 g) brown sugar
> ¼ cup (25 g) cocoa powder
> 1 tsp of vanilla extract
> ½ cup (640 g) whole wheat flour
> 1 tbsp cornstarch
> 2 tbsp psyllium
> ¼ cup (30 g) slivered almonds

Tip

Pinto beans, black beans, kidney beans and chickpeas can all be substituted for white pinto beans.

1. Preheat the oven to 400°F (205°C) and grease an 8 inch (20 cm) rectangular pan.
2. In a blender combine the plant sterol margarine, pinto beans, maple syrup and chocolate chips, blending until smooth.
3. Add the brown sugar, cocoa powder, vanilla extract, whole wheat flour, cornstarch, and pysllium and pulse for around 20–30 seconds.
4. Spread the mixture out evenly in a greased pan.
5. Top with slivered almonds.
6. Bake for 30–35 minutes or until fork comes out clean.

Nutrition Information: Fudgy Brownies			
Per Serving: 274 kcal and 5g of nuts			
Plant based protein	4.2 g	Total fat	13.1 g
Total carbohydrate	39.3 g	– Saturated	3.8 g
Dietary fiber	5.2 g	– Monounsaturated	4.1 g
– **Viscous fiber**	**2.1 g**	– Polyunsaturated	2.7 g
		Plant Sterols	0.9 g

GINGER PEACH CRUMBLE

Servings: 6
Time: 35–40 min

Filling
  2 ¼ cups (450 g) chopped peaches (about 4 medium sized peaches)

Topping

  2 tbsp almond meal
  ½ cup (50 g) oat bran
  3 tbsp rolled oats or oat bran

1 tbsp whole wheat flour
2 ½ tsp dried ginger
2 tbsp brown sugar
¼ cup (30 g) coarsely chopped almonds
Pinch of salt
¼ cup (55 g) plant sterol margarine

1. Preheat the oven to 350°F (175°C).
2. Grease an 8 inch (20 cm) pan and add peaches.
3. To make the topping, combine the almond meal, oat bran, rolled oats, whole wheat flour, dried ginger, brown sugar, coarsely chopped almonds and salt in a medium sized bowl.
4. Add the plant sterol margarine by cutting it into the topping mixture with two knifes. Stop when the texture of the topping resembles coarse gravel.
5. Sprinkle the topping peaches so they are covered. Bake for 30–35 minutes or until the peaches are soft and the topping is a golden brown color.

Nutrition Information: Ginger Peach Crumble			
Per Serving: 223 kcal and 5g of nuts			
Plant based protein	5.1 g	Total fat	13.0 g
Total carbohydrate	25.8 g	– Saturated	1.5 g
Dietary fiber	4.2 g	– Monounsaturated	5.4 g
– Viscous fiber	**1.0 g**	– Polyunsaturated	3.5 g
		Plant Sterols	0.9 g

BLUEBERRY LEMON CAKE

Servings: 12
Time: 45 min

Tip

You can substitute apple cider vinegar with any white vinegar.

Cake

1 cup (120 g) whole wheat flour
¾ cup (100 g) all-purpose flour

1 cup (200 g) white sugar
Pinch of salt
3 tbsp psyllium
1 tsp baking soda
½ tsp lemon zest
1 ½ tsp vanilla extract
⅓ cup (80 g) plant sterol margarine melted
1 cup (250 ml) soymilk
1 tbsp apple cider vinegar
1 ½ cups (150 g) frozen blueberries

Glaze

1 cup (130 g) icing sugar
2 tbsp lemon juice
½ tsp lemon zest
1 tsp. vanilla extract

Topping

⅓ cup (40 g) roughly chopped walnuts
¼ cup (25 g) blueberries

1. Preheat the oven to 350°F (175°C).
2. To make the cake, in a large bowl combine whole wheat flour, all purpose flour, sugar, salt, psyllium, and baking soda.
3. In a separate bowl mix lemon zest, vanilla extract, plant sterol margarine, and soy milk.
4. Add the wet ingredients to the dry ingredients, mixing sparingly.
5. Pour the mixture to a 25 cm (10 inch) Bundt pan.
6. Using a spatula fold in vinegar and blueberries.
7. Bake for 25–30 minutes or until fork inserted into the side of the cake comes out clean from batter.
8. While the cake is cooking make the glaze. In a small bowl combine the icing sugar, lemon juice, lemon zest and vanilla extract.
9. Let the cake cool for 10–15 minutes then drizzle glaze and add blueberries and walnuts to the top.

Nutrition information: Blueberry Lemon Cake
Per serving: 249 kcal and 3 g of nuts

Plant based protein	3.3 g	Total fat	7.9 g
Total carbohydrate	43.5 g	– Saturated	1.0 g
Dietary fiber	3.0 g	– Monounsaturated	2.5 g
– Viscous fiber	**1.2 g**	– Polyunsaturated	3.5 g
		Plant Sterols	0.6 g

CHOCOLATE MOUSSE

Serving: 8
Time: 3h

²/₃ cup (170 ml) aquafaba (the liquid from about 1 can (400g) unsalted chickpeas)
1 tsp lemon juice
2 packages (300 g) soft tofu
¹/₃ cup (60 g) semi-sweet (dark) chocolate chips, melted
2 tsp vanilla extract
1 cup (200 g) white sugar
4 tbsp cocoa powder

1. Drain chickpeas and collect the aquafaba liquid (about 170 ml) from the can in a large bowl. The chickpeas themselves will not be used in this recipe. For recipes using chickpeas see recipes in the other sections of this book, such as "Curried Hummus."
2. Add lemon juice and using an electric hand mixer, beat aquabafa until stiff peaks are formed (15–20 minutes). Do not over whip.
3. Melt chocolate using water bath. In order to do this:
   - Place chocolate in a small pot and nest the pot within a larger pot.
   - Fill a larger pot with just enough water so it is touches the bottom of the small pot, make sure not to get any water on the chocolate
   - Cook on medium heat stirring continuously, until chocolate has melted (around 5 minutes).
4. In a blender, combine the tofu with the melted chocolate, vanilla extract, sugar, and cocoa powder. Blend until smooth.
5. Combine the mixture in the blender with the aquafaba. Using a spatula, gently fold in blender mixture with aquafaba, adding the mixture slowly to make sure it becomes fully incorporated. Be careful not to over mix and fully deflate the aquafaba, although some reduction in volume is expected.
6. Pour into 6 serving containers (approx. ½ cup [125 ml] each).
7. Chill for at least 2 ½ hours or overnight for best results.

Tip

Instead of using the liquid from one can of chickpeas, the liquid used to cook dry chickpeas can also be substituted. About 1 cup of dry chickpeas cooked should yield the 2/3 cup (170ml) needed for this recipe.

Nutrition Information: Chocolate Mousse
Per Serving: 145 kcal

Plant based protein	6.5 g	Total fat	5.3 g
Total carbohydrate	19.5 g	– Saturated	1.5 g
Dietary fiber	1.6 g	– Monounsaturated	2.7 g
– Viscous fiber	**0.2 g**	– Polyunsaturated	0.6 g

ALMOND CLUSTER COOKIES

Serving: 22 cookies
Time: 1 h

1 ½ cups (170 g) slivered almonds
¾ cup (112 g) semisweet dark chocolate chips

Variation
Add 1tbsp of
orange zest to
melted chocolate.

1. Toast slivered almonds in a large pan over medium-to-low heat stirring occasionally until browned, around 5–10 minutes. Place toasted almonds in a large bowl and set aside.
2. Melt chocolate using a water bath. In order to do this:
   - Place chocolate in a small pot and nest the pot within a larger pot.
   - Fill a larger pot with just enough water so it is touches the bottom of the small pot, make sure not to get any water on the chocolate.
   - Cook on medium high heat stirring continuously until chocolate has melted, around 5 minutes.
3. Mix the almonds into the melted chocolate until the almonds are fully coated.
4. Scoop out a small spoonful of the chocolate coated almonds and place on wax paper lined baking tray.
5. Let cool for around 45 minutes in the fridge.

Nutrition Information: Almond Cluster Cookies
Per Serving: 68 kcal and 7g of nuts

Plant based protein	1.9 g	Total fat	5.1 g
Total carbohydrate	4.9 g	– Saturated	1.2 g
Dietary fiber	1.3 g	– Monounsaturated	2.3 g
– **Viscous fiber**	**0.1 g**	– Polyunsaturated	0.9 g

STRAWBARLEY PUDDING

A simple and fast treat!

> 1–⅓ cups (270 g) cooked barley (al dente)
> 2 tbsp. strawberry jam
> ½ cup (50 g) ground almonds (optional)
> 1 tbsp. sugar or 1 tsp. stevia
> Pinch of salt
> ¼ cup, or more soy beverage

1. Puree barley and jam together; adding enough soy beverage to make a pudding-like consistency.
2. Season with sugar, or sugar substitute.
3. Refrigerate until cool (approximately 2h). Garnish with jam, almonds or fresh strawberries.

Tip

For a creamier texture, add more soy beverage.

Makes 4 servings.

Nutrition Information			
Each 110 g serving made with almonds, 1/4 of the recipe contains:			
Energy	172 kcal	Monounsaturates	4.3 g
Protein	4.7 g	Saturates	0.7 g
Soy protein	**0.4 g**	Total carbohydrate	26.8 g
Total fat	6.9 g	Dietary fiber	4.1 g
Polyunsaturates	1.6 g	**Viscous fiber**	**1.3 g**

TANGY LEMON TART

Sweet and sour with a hidden tofu twist.

For the crust:
¼ cup (25 g) ground almonds
½ cup (50 g) oat bran
tbsp. almond butter
¼ tsp. salt
1 tbsp. water

For the filling:
2 cups (400 g) silken tofu
2 tsp. lemon zest (grated rind)
½ cup lemon juice (2 lemons)
2 tbsp. cornstarch
⅔ cup icing sugar

1. Blend together almonds, oat bran, salt and almond butter. Add water to help bind mix together.
2. Grease a 7" pie pan or tart pan. Press crust firmly onto bottom and sides of pan. Prebake crust at 350°F for 12 min.
3. To make the filling, pureé all ingredients in blender until smooth.
4. Pour filling into tart shell and bake at 350°F for 40 to 45 min (until brown at the edges and filling is set).

Makes 8 servings.
Recipe tested: Kathy Galbraith RD

Nutrition Information
Each slice, 1/8 of the recipe contains:

Energy	158 kcal	Monounsaturates	3.8 g
Protein	5.3 g	Saturates	0.8 g
Soy protein	**2.7 g**	Total carbohydrate	20.1 g
Total fat	6.9 g	Dietary fiber	1.4 g
Polyunsaturates	2.0 g	**Viscous fiber**	**0.5 g**

BARLEY RAISIN COOKIES

Skip the wheat and raise your fiber even higher when you bake with barley flour.

½ cup plant sterol margarine
½ cup brown sugar, packed
¼ cup granulated sugar
½ tsp. salt
½ tsp. cinnamon
¼ tsp. ground cloves
1 cup (120 g) barley flour
1 cup (90 g) rolled oats
1 tsp. psyllium
½ tsp. baking soda
½ cup sultana raisins
3 tbsp. water

1. Cream together the margarine and sugars.
2. Add spices, salt, barley flour, oats, psyllium, baking soda and raisins. Mix well. Stir in the water to moisten the dough, beat well.
3. Preheat oven to 350°F. On a greased baking pan, portion and flatten dough to form 20 cookies.
4. Bake for 10 min or until golden.

Tip

Try dried cranberries as a substitute for raisins.

Makes 20 cookies.
Recipe tested by Kathy Galbraith.

Nutrition Information
Each 30g cookie (1/20 recipe) contains:

Energy	107 kcal	Saturates	0.6 g
Protein	1.6 g	Total carbohydrate	15.8 g
Total fat	4.7 g	Dietary fiber	1.2 g
Polyunsaturates	1.8 g	**Viscous fiber**	**0.6 g**
Monounsaturates	1.8 g		

ALMOND, FRUIT FONDUE

Power up your day with fruit and the protein in almonds and soy.

2 cups (400 g) soft or silken tofu
150 mL soy yogurt or soy-based sour cream
1 tsp. almond extract
¼ cup (25 g) ground almonds
Maple syrup or stevia, to taste

1. In a blender, puree tofu with soy yogurt, almond flavor, ground almonds and sweetener.
2. Prepare a platter with the soy-filled fondue bowl. Surround with an assortment of fresh fruit for dipping.

Tip

Stevia is a naturally derived sweetener available at bulk, health stores and grocery health food sections.

Makes 4 servings

Nutrition Information			
Each serving contains:			
Energy	110 kcal	Monounsaturates	2.8 g
Protein	8.1 g	Saturates	0.7 g
Soy protein	**6.8 g**	Total carbohydrate	6.6 g
Total fat	6.1 g	Dietary fiber	0.8 g
Polyunsaturates	2.4 g		

Chapter 6

Why a Plant-Based Diet?

INTRODUCTION

It is undeniable that what we choose to eat affects not only our health and well-being but also the health of our planet. These individual choices are amplified by those of other people and have added up to a profound effect on the planet and the species that reside in it. In this way what we eat now is shaping the landscape of the world for future generations. Varying degrees of environmental harm are carried out depending on what is being produced, how it is being produced, and how far it needs to travel before being sold. Many things can be done in order to minimize the impact that food choices have on the environment. Of these, reducing consumption of animal products has the greatest effect. This is why plant-based or vegan eating patterns typically have the lowest environmental impact. Plant-based diets are also among the healthiest dietary patterns and have the added benefit of minimizing harm to other animals. The Dietary Portfolio was designed to be placed in the context of a plant-based diet for three reasons. The first being the strong evidence base supporting the health benefits of a plant-based diet, the second, its relation to sustainability, and the third, the ethical aspects of this type of dietary pattern. This chapter will outline how the Dietary Portfolio compares to other dietary patterns on each of these three factors, specifically those patterns that include animal products.

Key Points

- A large body of evidence exists linking the consumption of processed meats, red meat, and animal products high in saturated fat to cardiovascular disease (CVD) diabetes and certain types of cancer.
- Plant-based diets and diets low in animal products have been shown to reduce the risk of chronic diseases, such as CVD, diabetes, and obesity.
- Many dietary guidelines and health advisory groups now recommend either eliminating or limiting the consumption of processed meats, red meats, and other animal products high in saturated fat.
- Plant-based diets have a reduced environmental impact compared to those high in animal products. They are able to reduce:
  - Land use
  - Greenhouse gas emissions
  - Water use and contamination.
- Finally, ethical and moral questions surround the treatment of farm animals in this age of industrial agriculture. By relying on plant-based sources of protein such as nuts, soy, legumes, whole grains, and vegetables, the Dietary Portfolio minimizes it's contribution to harmful systems.

The Portfolio Diet for Cardiovascular Disease Risk Reduction. https://doi.org/10.1016/B978-0-12-810510-8.00006-6

A PLANT-BASED DIET AND HEALTH

For additional information: The Physicians Committee for Responsible Medicine is a nonprofit organization of over 12,000 physicians world-wide who are "dedicated to saving and improving human and animal lives through plant-based diets and ethical and effective scientific research." Visit their website at: https://www.pcrm.org/ for more information.

Following a more plant-based diet requires the restriction of animal products. Animal products include meats such as chicken, pork, beef, and fish as well as animal bi-products like eggs, milk, cheese, and butter. The exclusion of these foods has been practiced for over 500 years by Jain and Buddhist groups, who mainly consume a plant-based diet. Science is only now beginning to understand the many health benefits gained from this style of eating. Physicians are increasingly recommending a plant-based diet for health. Research has shown that eating plant-based diets can reduce your risk of chronic disease such as cardiovascular disease (CVD), diabetes, and certain types of cancers. This section outlines some of the landmark research in this area.

Meat Consumption and Chronic Disease

Red Meat

Diets high in plant-based protein have been shown to improve health outcomes compared with meat based diets.

Diets high in plant-based protein have been shown to improve health outcomes compared with meat-based diets, particularly those consisting of large amounts of red and processed meats [1–7]. In 2010, a Harvard research group lead by renowned nutrition researcher Dr. Walter Willett published one of the largest studies of its kind in relation to coronary heart disease (CHD) and meat consumption. The study collected data from 84,136 women in the Nurses' Health Study over the course of 26 years [1]. Red meat consumption was compared to other sources of protein such as nuts, beans, poultry, and fish. Risk of CHD was 13%–30% less when red meat was replaced with these alternative sources of protein [1]. The authors conclude that risk of CHD appears to increase with high levels of red meat consumption and shifting to alternate protein sources may significantly reduce the CHD incidence [1].

Red meat consumption has also been associated with increased CVD risk. Dr. Stanley Hazen of the Cleveland Clinic noted that trimethylamine-N-oxide (TMAO) occurs when substances (choline and carnitine) found in animal products, especially meats, undergo bacterial fermentation in the gut.

"...processed meat is carcinogenic to humans..." -International Agency for Research on Cancer.

Cell cultures studies have demonstrated that TMAO damages the vascular lining and increases inflammation [8]. It was therefore hypothesized that increased levels of TMAO would be associated with increased CVD. In a study published in the New England Journal of Medicine, it was demonstrated that an elevated level of TMAO predicted an increased risk of major adverse cardiovascular events during three years of follow-up in patients with existing CVD [9]. TMAO has also been described as a potential target for insulin resistance [10], and studies have shown an increased risk of diabetes in individuals with higher TMAO levels [11].

Researchers have also found significant evidence that red meat consumption during adolescence is linked to premenopausal breast cancer. In a study consisting of 44,231 women, those who consumed the most meat during adolescence had a significant (43%) increase in breast cancer compared to those who consumed the least [12].

The International Agency for Research on Cancer (IARC) reported that "...processed meat is carcinogenic to humans [...] and red meat is probably carcinogenic to humans" and may be linked to colorectal cancer [13]. The conclusions of the IARC were based on evidence collected from over 800 studies, across 10 countries, in which processed meat was reported to show convincing evidence of causing colorectal cancer [13]. In addition to colorectal cancer, positive associations have been found for the consumption of processed meat and stomach cancer, as well as red meat and pancreatic and prostate cancer [13].

The connection between meat consumption and negative health outcomes is reflected in the dietary recommendations by numerous international health authorities (Table 6.1) and an increasing amount of dietary guidelines provided by individual countries. The dietary guidelines in America recommend a vegetarian diet as one of the three recommended healthy eating patterns [24]. The Canadian guidelines recommend individuals "Choose protein foods that come from plants more often" [25]. The Netherlands call to limit meat consumption to no more than two servings per week [26], and Belgium suggests making foods derived from plants the center of each meal and to limit animal product intake [27]. Chinese guidelines call for a reduction in meat consumption by 50% [28].

TABLE 6.1. International Health Authorities Dietary Recommendations

Year	Organization	Recommendations Regarding Meat Consumption
2018	The World Health Organization Europe [14]	• Eat foods which consist of "mainly plants rather than animals."
2015	The World Health Organization [15]	• Limit consumption of fatty meat. • Increase consumption of fruits and vegetables, and legumes, whole grains, and nuts.
2009	European Food Information Council [16]	• Limit the consumption of animal products containing high total and saturated fat.
2017	American Diabetes Association [17]	• Does not recommend red meat consumption.
2017	Diabetes Canada [18]	• If meat is consumed, fish, and lean meats are advised.
2016	Canadian Cardiovascular Society [19]	• Emphasizes a diet rich in vegetables, fruit, whole-grain cereals, and polyunsaturated and monounsaturated oils. • Recommends the Dietary Portfolio.
2017	HEART UK [20]	• Consume less fatty meats.
2016	European Atherosclerosis Society [21]	• Advocates for a more plant-based diet. • Recommends the Dietary Portfolio.
2015	The International Agency for Research on Cancer [22]	• Limit intake of red meat.
2014	American Heart Association [23]	• Reduce meat consumption. • Limit fatty/processed meats.

Fish

Fish are widely promoted by health authorities as good source of protein and healthy fats. The primary reason for this being the presence of high levels of omega-3 fatty acids (eicosapentaenoic acid and docosahexaenoic acid) found in fish which have been purported to reduce the risk of CVD [20,24,29]. However, scientific opinion is divided. A Cochrane review analyzed over 15,000 publications on the topic and found no significant protective effect on total mortality from cardiovascular events or cancer [30]. A recent literature review supports this finding showing no consistent protective effect of fish oil supplementation on CVD [31]. Positive results have undeniably been seen in cohort studies, where populations are followed over time; although conclusions can be difficult to draw from studies of this nature [32,33]. For example, people who eat fish tend to be more health conscious, do not smoke, and have healthier overall diets [34–36]. As a result of these confounding factors a clear causal link cannot be made between fish consumption and health. In addition, increasing levels of water contaminants such as mercury, dioxins, and PCBs, that may be detrimental to human health, are accumulating in fish [37]. Additionally, some research has shown that the consumption of long-chain n-3 fatty acid supplements derived from fish, are not beneficial for men with angina [38].

> A Cochrane review, analyzed over 15,000 publications on the topic finding no significant effect on total mortality or death from cardiovascular events or cancer [30].

Overall, it appears that recommendations promoting fish consumption have paid too much attention to studies showing benefits and have failed to consider all the emerging evidence on risk. More work is needed looking into both fish and fish oil supplements to determine whether there truly are benefits to consumption. These results should also be compared with alternatives, such as algae oil supplements which also contain long-chain n-3 fatty acids if strong conclusions about the benefits of fish consumption are to be made.

Animal Bi-products Good, Bad or Somewhere in Between?

Dairy

Milk is a rich source of nutrients and is widely promoted for bone health and the prevention of bone density-related fractures. Many countries' dietary guidelines typically promote milk consumption throughout the human lifespan. However, the evidence behind this may not be so clear cut. Two large observational cohort studies in Sweden including a total of 106,772 individuals, conducted over the course of 20 years, found that dairy consumption may cause more harm than good [39]. Specifically, they found that increased milk consumption was associated with increased overall mortality and bone density-related hip fractures in women. These findings are believed to be the result of D-galactose, a type of sugar of which milk is the primary dietary source [39]. D-galactose has been shown to induce the aging process in animal studies, inducing oxidative stress damage, inflammation, neurodegeneration, decreased immune response, gene transcriptional changes, and other factors related to shortened lifespan, CVD, cancer, bone loss, and sarcopenia [40,41]. Adding to this body of evidence, findings from the Nurses' Health Study indicate that milk consumption in the prepubertal period was related to an increase in hip fractures in later life [42].

> Two large observational cohort studies in Sweden found an increase in overall mortality or death in addition to an increases in bone density related hip fractures in women associated with high levels of milk consumption.

In contrast, a recent meta-analysis which included the Swedish cohort study, found no strong associations, positive or negative, between all-cause mortality, CVD and milk consumption [43]. Despite these results the authors conclude that risk cannot be fully dismissed due to possible publication bias as papers showing no effect are rarely published.

An association between milk and certain types of cancers has also been proposed [44,45]. Milk consumption stimulates the production of a compound called insulin-like growth factor (IGF-1) which has been implicated in the development of reproductive cancers [44,45]. A study found men with the highest levels of IGF-1 to be four times more likely to develop prostate cancer [46]. This is in line with an observational study conducted in Japan which found a 20-fold increase in dairy consumption to be accompanied by the fastest growing rate of prostate cancer in the world [47]. On the other hand, dairy consumption, possibly due to its calcium content, has been related to reduced rates of colon cancer [48].

While these studies provide a thought-provoking perspective on the role of dairy consumption and health more research is needed in to assemble a clearer picture and determine whether milk substitutes such as soy or almond, could more favorably fill the dietary niche that milk has occupied.

> In the Japanese population, a 20-fold increase in dairy consumption was associated with the fastest growing incidence rate of prostate cancer in the world.

Eggs

Eggs are another food typically perceived to be healthy. As is the case with milk products, the scientific community has differing perspectives on its effects on health. This debate has largely centered around the fact that eggs are a significant source of dietary cholesterol, with one egg yolk containing an amount almost equivalent to a 12-ounce hamburger [49]. It was originally suggested that consuming a high cholesterol diet would increase an individual's own serum cholesterol level. As high levels of low density lipoprotein cholesterol (LDL-C) are a risk factor for heart disease, it was recommended that foods high in cholesterol be avoided. However, recent research has found that for most people the dietary cholesterol found in eggs will not cause their own cholesterol levels to rise significantly [50–52]. Although for a small percentage of people the amount of dietary cholesterol consumed from sources like eggs may increase cholesterol levels [53]. This effect is a result of a genetic variation that makes individuals more sensitive to dietary cholesterol. For these individuals the high amount of dietary cholesterol found in eggs is potentially harmful [53]. However, even if individuals do not possess this genetic predisposition, research has found that for people at risk for CVD, egg consumption can still be detrimental due to the production of TMAO when choline in eggs is fermented in the gut (see Meat Consumption and Chronic Disease for a discussion on TMAO and its link with CVD and cancer) [49].

In addition, for individuals with diabetes, the consumption of eggs has been associated with an increased risk of heart disease [54]. The consumption of eggs was also found to be associated with an increased incidence of type 2 diabetes [54]. These results suggest that for most individual's moderate egg consumption may have no effect on cholesterol. However, a recent population study demonstrated overall adverse effects of egg consumption for individuals with certain risk factors [55]. Therefore, for certain groups like diabetics, individuals at risk of CVD, or those with a specific genetic variation, the consumption of eggs should be avoided.

> In individuals with diabetes, the consumption of eggs has been associated with an increased risk of heart disease.

PLANT-BASED DIETS AND THE PLANET

Reducing the impact humans have on the environment is arguably the most important issue of our time [56,57]. Our actions are not only shaping the world for future generations but are increasingly affecting those living today. Climate change and environmental degradation are causing land to be swallowed up by the rising tides from melting polar caps and glaciers. Food shortages are increasing due to mass drought and species extinction is on par with that in the Cretaceous period which wiped out the dinosaurs.

Climate change is occurring at such a rapid pace that it is visible in our lifetime. In Canada, climate change can be seen in occurences such as the disappearance of the glaciers. Canada is home to some of the most visited glaciers in North America, admired for their iconic grandeur and the exciting habitats created as the result of them. In Alberta, the Columbian Icefields are declining at a rate of over 5 m/year and are expected to disappear completely over the next decade. Beyond their function as a popular tourist destination and home to various creatures, the deterioration of the glaciers has implications for hydropower generation and water supply systems to lower regions, as melt and thaw cycles affect water flow. Damage to irreplaceable ecosystems services, like those provided by glaciers, is increasingly occurring all over the world. We live in a pivotal time to prevent these changes and curb the effects climate change will have on both people and the planet. One of the major ways that individuals can reduce their environmental impact is through dietary change.

Agriculture and the Environment

For those fortunate enough, food is a regular part of daily life. However, how much do we really know about where our food comes from? In the grocery store we are disconnected from the reality of what growing fruits and vegetables and raising animals entails. A trip around the store shows off fruits and vegetables grown from seeds you never planted, in soil you never set foot on. Dairy and meat products are conveniently arranged in neatly shaped bottles and packages unrecognizable as the animals from which they came. The distance many of us have from the products we consume makes it easy to forget the enormous work and tremendous amount of resources that go into making a meal. Even though this process may seem far removed from our lives, we are very much connected to the consequences. By understanding the effects of our food choices it may be easier to pick products, not only for health and taste, but also to reduce environmental impact. Some of the main environmental impacts of the food system are as follows:

Water use and contamination: Marine ecosystems are heavily polluted by agriculture. Pesticides, herbicides, fertilizers, and manure are washed away from fields and deposited in waterways causing contamination. Fertilizers and manure serve to feed massive algal blooms, depriving other aquatic inhabitants of oxygen with their rapid and extensive growth. Some algae produce toxins that are poisonous to people and other animals. Once contamination has occurred, algal blooms are extremely costly and difficult to control.

In Canada, Lake Erie is a famous example of the problems that an over abundance of nutrients can present. In the 1970s pollution was so bad that Dr. Seuss included the following line in the original copy of the Lorax, a well-known children's book "They'll walk on their fins and get woefully weary, in search of some water that isn't so smeary. I hear things are just as bad up in Lake Erie." Although great improvements have been made, agricultural runoff continues to be a problem. To this day Lake Erie is still plagued by algal blooms as agriculture continues to encroach on this region, making the future of this lake uncertain [58]. In addition to contamination issues, the volume of water used by agriculture is extremely high. This high usage is becoming increasing problematic as climate change has exacerbated drought in many regions. In a time when fresh water is becoming scare, preserving the few fresh water sources we have is a global priority.

"They'll walk on their fins and get woefully weary, in search of some water that isn't so smeary. I hear things are just as bad up in Lake Erie."
—Dr. Seuss

In terms of agricultural water use, plant-based foods require substantially less than animal products. This observation is even true when protein content is taken into account. Fig. 6.1 illustrates that per gram of protein, milk, chicken, and eggs use around one and a half times more water compared to pulses (legumes and beans) [59]. The greatest difference can be seen between cattle and pulse crops, with cattle requiring six times the amount of water per gram of protein compared to pulses [60,61]. While it is true that unlike the animal products they are being compared to, pulses do not have a complete source of protein (as discussed in Chapter 4 under tips for eating a plant-based diet), it is possible to combine pulses with other plant-based foods to form a complete source of protein that still uses less water than cattle production. For example, pulses, such as kidney beans or chickpeas, can be combined with a grain, like rice, to form a complete protein that still requires less water to produce than cattle. This process can be used with many other plant-based protein sources to create complete proteins that use less water to produce than animal-based sources (for more combinations see Chapter 4). It also is important to note that protein deficiency is incredibly rare in developed countries and a plant-based diet high in fruits, vegetables, legumes, and whole grains will provide more than sufficient protein for most individuals.

Greenhouse gas production: Greenhouse gases (GHGs) are atmospheric gases which act to prevent heat from escaping. If you have ever walked in a greenhouse, you may have noticed it is warmer than the surrounding area. This difference is because the glass allows heat to enter but stops some from exiting, allowing heat to accumulate. The "greenhouse gas effect" works in much the same way. When heat from the sun enters our atmosphere it is trapped on the surface of our planet by GHGs. The abundance of GHGs in our atmosphere has resulted in the temperature of the planet and its oceans rising. However, warmer temperatures are not the only consequence of increased GHGs. Climate change induces changes in precipitation and wind patterns, which in combination with rising temperatures is incrementally altering the systems which form everyday weather. On our current trajectory we will not meet emission targets set out by expert organizations to minimize the negative effects of climate change [62]. Our failure to meet these targets could have devastating consequences for people and other life forms on our planet.

Around 30 years from now GHG emissions from agriculture and land clearing are predicted to increase by a staggering 80%. An increase equivalent to the GHG production of the entire 2010 transportation industry.

FIG. 6.1   Water use per gram of protein [63].

Currently, food production is responsible for around one-quarter of global GHG emissions [64]. In 30 years emissions from agriculture and land clearing are predicted to increase by a staggering 80% [65]. An increase on this scale is equivalent to the GHG production of the entire 2010 transportation industry [66]. Emissions from food production come from a variety of sources. Before food leaves the farm, emissions are mainly the result of the application of fertilizers, production of livestock, and combustion of fossil fuels used to power machinery like plows and tractors. After food is produced, a whole new set of emissions come from the transportation, processing, storage, preparation, and waste disposal of food. While the amount of emissions vary by region, meat and dairy products are consistently the biggest sources of GHGs in the food sector irrespective of production methods [67] (Box 6.1).

Compared to other diets, plant-based diets have the highest potential for reducing total per-capita GHG emissions (up to 20%) [71]. The large emission reductions seen on a plant-based diet are predominately owing to the elimination of animal products. Cattle account for 14% of human-related GHG emissions [61]. Through the digestive process, cattle and other ruminant animals naturally produce methane, a GHG which has 25 times the heat trapping ability of carbon dioxide. Cattle production also contributes to emissions from the clearing of land for pasture, particularly when forests are burned, globally producing billions of tons of carbon dioxide each year.

> Compared to other diets, plant-based diets have the highest potential for reducing per-capita GHG emissions.

Land use: By changing the landscape into areas for growing crops or raising animals we are not only changing ecosystems but also the capacity of an area to remove or sequester carbon dioxide from the atmosphere. This effect is particularly

---

BOX 6.1 Grass-Fed vs. Grain-Fed Cattle

Grain-fed cattle typically reside in concentrated feedlots or industrialized systems where feed is grown separately. Grass-fed cattle are raised on a pasture consisting of grasses and sometimes woody shrubs and trees. Industrial systems sometimes finish the animal on a pasture but for most of its life it will have been fed food grown elsewhere.

Contrary to popular belief, consuming grass-fed cattle does not decrease environmental impact compared to grain fed cattle [68]. Cattle raised in systems where they are given grain instead of grass typically produce less GHG emissions and use less energy per pound of product compared to those raised on pasture. The energy dense nature of the grain feed allows cattle to grow much faster and therefore they are ready for slaughter sooner [69]. In other words, the overall emissions cows produce throughout a lifetime in a pasture system are greater than those that go into feeding and housing cattle in industrial systems as these cattle have shorter lifespans. However, industrial systems have a host of other issues such as the creation of concentrated sources of manure and waste, antibiotic use as well as welfare concerns (discussed under Ethical Considerations). There is an argument to be made that marginal lands which cannot grow other crops, produce a net gain by having cattle grazing on them, as in these instances this approach is the most efficient use of space. While it may be true this method of farming maximizes food yield in a given area, it will still produce large quantities of GHG emissions. Although the shrubs and grasses in a pasture system can remove carbon dioxide from the atmosphere, the emissions produced by grazing cattle will be greater than the emissions the pasture can absorb [70]. In addition, this argument does not take into account that marinal lands are the habitat of other, sometimes threatened species or that these species may be essential to the health of the environment.

apparent in areas where deforestation has occurred. While all plants are able to trap carbon dioxide in tissue, trees are better able to do so owing to their long lives and large woody trunks and branches.

Throughout the world many forest ecosystems are being converted to agricultural land resulting in a poorer capacity for carbon sequestration. Agriculturally driven deforestation is predominately happening in South and Central America, with forested areas in Central America reduced by 40% in the past 40 years [72]. Approximately two-thirds of the deforested lands are converted to pasture [72]. For other countries with limited convertible land, the increasing demand for livestock has led to the reliance on industrialized systems, with feed for livestock often imported from elsewhere [72]. However, importing feed shifts the environmental burden of feed production from one country or region to another, extending the food production capacity of one region at the expense of another.

Eating a plant-based diet has been shown to have the highest potential for reducing land use (reductions up to 60%) compared to other diets [71]. By far, the most extensive amount of land used by the agricultural industry is designated for cattle production. 26% of the earth's ice free area is covered by livestock grazing operations and 33% of cropland is used for livestock feed [73]. The surface area necessary to raise cattle is increasing in response to growing demand [73].

> 26% of the earth's free area is covered by livestock grazing operations.

Using the finite amount of land on this planet to rear animals is inefficient when crops suitable for human consumption are displaced by crops for livestock feed. Most of the energy gained from feed is lost in the animal's bodily functions such as keeping warm and breathing. Typically, not more than one-third and sometimes less than one-tenth of the energy value of the food that is fed to an animal is conveyed back to the consumer [74]. In other words, we are putting in more resources than we are getting out. The food that is designated for feed is often high quality and suitable for human consumption. If we are to preserve any wilderness while simultaneously creating a food system able to feed the 9 billion people expected by 2050, efficient use of space is of the upmost importance.

Species loss: We share this planet with a diverse range of unusual and spectacular creatures. Habitat loss and degradation is the greatest driver of species extinction, globally accounting for 84% of the species on at risk listings [75].

In Canada, like many other countries, agriculture has the most profound effect on species endangerment. Around 70% of wet lands have been drained and of those remaining 60%–80% are significantly polluted with fertilizer and pesticides from agricultural activity, effectively destroying them as a habitat for many species that live there. This loss is particularly significant as Canada is estimated to house one-quarter of the planets remaining wetlands [76] Canadian forests have also come under threat as a result of expanding cities and the subsequent need for increased food production. Carolinian forests in southwestern Ontario are the most biodiverse areas in the country supporting 40% of the breeding bird population. This area has already largely been taken over by agricultural activities and urbanization leaving less than 5% of the original woodlands. Species loss from human activity, as illustrated by Canada, is a global phenomenon occurring to varying degrees throughout the world (Box 6.2).

> Habitat loss and degradation is the greatest driver of species extinction, globally accounting for 84% of species at risk listings.

BOX 6.2 A Closer Look at Species Loss

Over the past 50 years, species extinction has occurred at a rate unprecedented in human history and to an extent unseen for the past 75 million years, since the extinction of the dinosaurs. This mass extinction has led many scientists to characterize the modern epoch as the Anthropocene from the Greek "Anthropos" meaning "human being." The Anthropocene is characterized by a time of mass extinction as the result of human activities. The "Red List" documents these extinctions and classifies animals based on the level of threat they are experiencing (http://www.iucnredlist.org/). Species range from least concern, meaning they are abundant and healthy, to extinct. Threatened species exist in a category in between these two extremes where animals are vulnerable, endangered, or critically endangered.

Globally, human activity has resulted in habitat loss and habitat degradation accounting for:

- 91% of threatened plants
- 89% of threatened birds
- 83% of threatened mammals

Even protected areas cannot escape the reach of human influence, particularly when the animals that inhabit them do not understand the new boundaries determined for them. Despite conservation efforts, turtle populations around Point Pelee in southern Ontario, Canada have shown a rapid decline. Since the time these ancient animals have existed, they have evolved intricate strategies for survival. The Blanding's turtle (*Emydoidea* blandingii) is an extremely long-lived species for an animal of its size, with the ability to live past the age of 75. Uniquely, it has a hinged shell that it can fully close over its feet and head to ward off any unwelcome visitors. These adaptive strategies have served it well up until roads were built around its nesting areas and pesticides polluted the water in which it lives. Now it is unable to adapt to the fast changes occurring around it and its survival is threatened. Many extinction events like this are occurring globally as human activity drastically changes the environment. It is a great shame that what has taken hundreds of millions of years to evolve is disappearing in generations.

Dietary Guidelines

The dietary guidelines of a country are taught as part of school curricula, influence federal policy, and form the building blocks for diet-related programs such as school lunches. In response to the urgent need to improve the environmental impact of diet, guidelines around the world have begun to include sustainability as a criterion for making dietary choice. When taking a long-term view, it is clear that health and sustainability are intimately linked. At its core, sustainability aims to ensure future generations have the same quality of life, if not better, than those living currently. Continuing to degrade the land, pollute the water and fuel climate change with our agricultural emissions will not be possible if we plan to provide our children and our children's children with the same healthy options many of us are lucky enough to have today. Given the huge impact diet has on the environment, it stands to reason that sustainability should be a consideration when formulating dietary advice for the public.

As of 2018, the Netherlands, Sweden, Qatar, Brazil, the United Kingdom, and Germany have all included sustainability in their guidelines. Progressively guidelines are beginning to include sustainability as part of dietary advice. This seemingly slow transition is in part due to the nature of dietary guidelines, which makes them slow to adopt new recommendations. In many countries new guidelines are only released every 5 or so years, which has the benefit of insulating guidelines from passing trends and allowing time for evidence to accumulate. However, in regard to the link between diet and sustainability, the quantity and quality of evidence has not been the issue. In this instance, the main resistance has not been from the scientific community but instead from lobbyist groups for whom these recommendations threaten the survival of their associated industries. Sweden and the United States are examples of where lobbyist groups have opposed the inclusion of sustainability in dietary guidelines, with differing results. The current Swedish guidelines include evaluations of sustainability, while the American guidelines do not. Although Sweden is now thought to have one of the most progressive sustainability frameworks within their guidelines, they faced substantial struggles initially.

> The Netherlands, Sweden, Qatar, Brazil, the United Kingdom and Germany have all adopted sustainability into official dietary guidelines.

Sweden

The first attempt to include sustainability in Swedish guidelines was in 2008, but it was not until 2014 that they were accepted. This delay has been attributed mainly to the influence of industry, particularly those industries with large environmental impacts, like the meat producers. The current Swedish guidelines include evaluations of the consequences of meat consumption: "Of all foods, meat has the greatest impact on our climate and environment. This is why it's important for us to cut back on meat and be careful about what meat we do choose to eat." From these types of statements, it is clear why the meat industry would take exception to sustainability being included in the guidelines. Dairy producers had similar objections towards about the guidelines' statements on limiting consumption of dairy products for environmental reasons as this was thought to reduce dairy consumption. Dairy boards petitioned to have this segment removed on the grounds that people would not receive enough calcium. However, this concern was rejected by the dietary guidelines committee and instead an appendix was added on high calcium plant foods, of which there are many.

> "Of all foods, meat has the greatest impact on our climate and environment. This is why it's important for us to cut back on meat and be careful about what meat we do choose to eat."
> —Swedish Dietary Guidelines, 2014

United States

Countries with stronger lobbing groups, like the United States, have been unable to overcome the influence of special interest groups in creating dietary guidelines. Sustainability is not mentioned in the American

Dietary Guidelines despite being a main theme in the America Dietary Guidelines Advisory Report, the preliminary report which compiled the evidence base for the dietary guidelines. Even purely health-related claims in regard to meat consumption were severely toned down with recommendations in the advisory report calling for decreased consumption of red meat and in the final report calling for increased consumption of lean meat. This aspect of the current American Dietary guidelines can be seen as a failure on the part of the United States to base policy on science rather than politics. As a result, many health practitioners have refused to use the current American dietary guidelines instead opting to use the preliminary report. Dr. David Katz, a world-renowned nutrition researcher and founding director of the Yale Griffin Prevention Research Center, succinctly sums up the situation "We are awash in preventable chronic disease. We are eating away our own health. We are eating our children's health, and their food, and drinking up their water. We are, into the bargain, devouring our very planet. Yet we are told here to keep on keeping on. That's what you get when it is politics, rather than science, on the plate." In a time when politics determine government policy it is increasingly important to use the available scientific information to make informed choices for yourself. Even if you live in a country that does not include sustainability in their guidelines there are many things you can do to reduce the environmental impact of your diet.

"We are awash in preventable chronic disease. We are eating away our own health. We are eating our children's health, and their food, and drinking up their water. We are, into the bargain, devouring our very planet. Yet we are told here to keep on keeping on. That's what you get when it is politics, rather than science, on the plate."
—Dr. David Katz

What Can You Do?

Luckily when it comes to food, healthy diets overlap greatly with those diets that are less harmful to the environment. Eating a plant-based diet high in nuts, fruits, vegetables, and legumes like the Dietary Portfolio is one way to accomplish better health and environmental outcomes. While changing diet alone will not solve climate change, it is still one step in the right direction. Unlike many other steps that must be taken, changing your diet does not rely on government agencies or corporations to take action. In this way individuals have some autonomy over their own environmental impact. How to follow the Dietary Portfolio in the most sustainable way is discussed below.

Eat Plant-Based Foods

- Eat Portfolio foods in the context of a plant-based diet.
- Decrease or eliminate the consumption of foods from animal origin such as beef, dairy, pork, chicken, eggs, and fish.

The Dietary Portfolio is set in the context of a plant-based diet. Plant-based diets include all foods that are not of animal origin. Foods containing animal bi-products such as butter, milk, and eggs are all considered animal products. The goal of a plant-based diet is to completely eliminate these products. However, this can be difficult and for some it may be more helpful to view a plant-based diet as a direction to walk, with the path being as wide or narrow as you choose. Dietary change has the most potential for reducing the environmental

Additional Information
When eating out use search engines like Happy Cow to find plant-based restaurants in cities across the world: https://www.happycow.net.

impact of your diet [77]. Compared to other eating patterns, plant-based diets have the highest potential for reducing per-capita GHG emissions, land use, and water use. The more plant-based foods you eat and the less animal products, the smaller your environmental impact will be. Following the Dietary Portfolio in the context of a plant-based diet will maximize both health and environmental benefits (Box 6.3).

Buy Local

- Where possible, consume locally grown Portfolio foods.
- Look for locally grown products at grocery stores and farmers' markets.
- Avoid greenhouse grown foods by shopping seasonally when possible (seasonal guides can be found online for your region).
- Investigate where your soy products come from. Try to avoid areas where mass deforestation is occurring like in Amazon rainforest.

For most food, the fuel required for transportation makes up a relatively minor portion of overall GHG emissions [70]. However, food transport allows for the consumption of foods beyond what the land would naturally be able to produce, as local produce is supplemented with imported [70]. In this sense importing food is essentially exporting environmental degradation, as those countries must now deal with decreased natural spaces, soil erosion, water pollution, and other associated impacts. This issue can be particularly problematic in instances where rich countries rely heavily on poorer countries to produce animal feed, distorting local economies, and using up local resources [70]. In addition, while the total emissions from food imports are relatively small compared to other sectors in the food system, as the population increases so will emissions, making these currently minimal emissions significant [70]. Buying fresh produce at local grocery stores or farmers' markets will help to ensure transport costs are kept low. However, it is

---

BOX 6.3 Fish and Sustainability

Recommendations to increase consumption of fish pose a seeming contradiction between health and sustainability. However as discussed in "A Plant Based Diet and Health," the health benefits of fish consumption may not be as substantial as commonly believed. What is clear is that increasing demand is causing an extinction crisis. Fish have the largest total number of threatened species, with the Mediterranean alone having 56% of its endemic fish species threatened with extinction [63]. Global fish catches have been in decline since the 1950s resulting in exponential fishery collapse [78,79]. In Canada for example, overfishing famously resulted in the cod fishery collapse and mass unemployment [80]. Scarcity has not quenched demand and instead has fueled a large import market for fish, resulting in the disruption of food systems in poorer countries [81]. This new market has inflated prices in these countries, turning what once was a staple food into luxury goods [81].

Fish farms have been proposed as a sustainable alternative to wild caught fish, however these are not without their own set of issues. Farmed fish are typically kept in containment within natural ecosystems. Offshore fisheries have been linked with the spread of disease and parasites as well as water contamination from waste [82]. Lager carnivorous fish such as salmon and Bluefin tuna are typically fed small fish caught in the wild and continue to put pressure on wild fish stocks [83]. Typically, 2.5–5 kg of food is needed to produce 1 kg of farmed carnivorous fish [83]. Like the other industrial systems discussed in this chapter, the energy in is less than what comes out. Overall, neither farmed fish nor wild caught fish can provide the environmental protection necessary to ensure a sustainable solution. The current health claims made for fish and the accompanying recommendation for consumption may not be sufficient to justify the environmental damage fish consumption is causing.

important to note that foods grown in greenhouses, even when local, often use large amounts of energy for heating and other maintenance. To minimize environmental impact when buying greenhouse grown produce, look for some form of environmental certification stating that the impact of the greenhouse was reduced through incentives such as sustainable power generation [84]. In some instances, it may be more beneficial to consume imported products than those from greenhouses.

Following the Dietary Portfolio in the context of a plant-based diet necessitates the consumption of large quantities of fresh fruits and vegetables. Plant-based diets have a lot of flexibility in terms of what to buy making them suitable for many different regions. While the majority of Dietary Portfolio foods are available locally in various regions, some viscous fiber foods such as okra, eggplant and psyllium are limited to smaller growing regions and may not be available locally [85]. While the emissions produced by growing and transporting these foods is still lower than growing and transporting animal products, importing foods does transfer the burden of production to other countries. When possible, try and find local sources of these ingredients.

Avoid Heavily Processed and Packaged Food

- Buy whole foods, focusing on a diet rich in fruits, vegetables, and legumes.
- If you enjoy meat analogs such as soy burgers or mock chicken, try alternating between legumes and meat analogs as main protein sources for each meal.
- Avoid prepackaged fruits and vegetables.
- Bring your own bag when shopping.

Avoiding heavily processed foods, particularly when large amounts of salt and sugar have been added, can benefit both your own health and the environment. The environmental impact of processed foods varies between products and depends on the degree to which they must be processed, as well as the power source used. If the energy used for processing comes from wind or solar power, products will be less likely to have a large environmental impact than if they are generated through more traditional methods, like burning fossil fuels. That being said, it is not always clear when walking around a grocery store where products are produced and what the associated energy costs will be for processed foods.

The Dietary Portfolio uses many whole foods such as nuts, pulses, whole grains, fruits, and vegetables, which require low amounts of processing. Many of these foods can be bought with a minimal amount of packaging, especially dry foods like nuts, barley, psyllium, and pulses, which can be obtained from bulk stores. The Portfolio diet also contains a small amount of processed items including tofu, soy milk, soy meat analogs, and plant sterol margarine. While avoiding plastic packaging is more challenging for these items, the energy used to make these products is still less than their animal-based equivalents. In studies comparing chicken, the lowest impact meat, with a variety of plant-based substitutes, soy products had the lowest environmental impact aside from pulses [86] (Box 6.4).

Soy milk is also a staple in the Dietary Portfolio and although it does require processing, compared to dairy alternatives, it is still less harmful to the environment. In preprocessing, soy milk has many environmental advantages over dairy including reduced land usage, less emissions per liter of product, and smaller water requirements, all of which lower its relative impact [87]. After appropriate long-life processing, soy milk may not require refrigeration, lowering its energy usage, and extending its shelf life.

In the future it is likely that the sustainability of meat and dairy substitutes will be pushed even further. Initiatives such as "Plant Meat Matters" at Wageningen University are looking to optimize meat analogs,

BOX 6.4 Does Tofu Cause Deforestation?

Soybean farming is rapidly expanding, particularly in regions like South America [72]. This expansion has led to large-scale deforestation [72]. As soy products such as tofu, tempeh, and soymilk are a staple of many plant-based diets, including the Dietary Portfolio, there is a growing concern as to where to obtain the soy. However, the majority of soy grown today is fed to animals [72]. It is important to note that the expansion of the soy industry is not because of growing demand for soy products for human consumption but instead the result of the growing demand for animal feed [72]. If grown sustainably, soy products are a great option for environmentally conscious consumers. Compared to milk they require less land use, generate fewer GHG emissions, and require less water use. Some brands which are actively engaged in providing sustainable soy products are Alpro®, Eden Soy®, Sunrise Soya Foods®, Provamel®. Try searching brands on websites such as: http://www.ethicalconsumer.org/buyersguides to find out more about the products you consume.

not only for a more comparable texture, but also to minimize the environmental impacts of their production. In addition, industry is becoming increasingly aware of the growing demand for sustainable products. Impossible Foods™ developed by Stanford professor emeritus, Dr. Patrick Brown produces sustainability driven plant-based burgers. In addition to being hailed as the closest thing to meat on the market, Impossible Foods™ strives to continually reduce the environmental impact of its products by focusing on sustainable sourcing, minimizing water use, reducing emissions, and waste. The ambitious end goal being zero waste in the future. Alpro®, a large dairy analog company which produces soy milk, soy yogurt, etc. also includes sustainability as part of the company mandate. These goals include sourcing soy locally as well as reducing overall emissions and water use.

Reduce Food Waste

- Do small grocery shops weekly or daily.
- Plan meals ahead of time.
- Buy hardy fruits and vegetables such as dried legumes, apples, and members of the brassica family (cabbage, broccoli, cauliflower) as they are less likely to spoil before you can eat them.
- Buy perishable foods such as lettuce, fresh herbs, and berries in quantities that can be eaten over a few days.

Globally one-third of all food produced gets wasted. Wasting food means all the GHGs, land use, and water use that went into making these foods was essentially for nothing. Much of this waste is largely unavoidable and happens during harvest. However, in richer countries such as North America and Europe over 40% of food waste happens at the consumer stage [88].

In order to minimize food waste at home many strategies can be employed. Freezing foods has relatively minimal energy costs and can be a useful strategy for extending the life of leftovers or perishables. Small, frequent, grocery trips and planning meals in advance can also help to reduce foods waste for perishable items. Efficient food planning may involve buying foods for certain recipes or more generally, buying short-lived items such as lettuce, berries, and fresh herbs in quantities that you can eat over the next couple days and buying larger quantities of long lived vegetables and fruits such as apples, oranges, brassica (cabbage, broccoli, cauliflower, etc.), and dried legumes to eat later on.

Creating a Green Future

Resources on this planet are finite and current consumption patterns are on a trajectory of exceeding planetary boundaries. We are using resources at a rate that exceeds their ability to regenerate. Undoubtedly, technological advancement will help solve these problems. However, these advances will be unable to elicit sufficient change in the time frame that is needed. What you purchase on a day-to-day basis is one of the ways you can have the greatest impact on the health and happiness of future generations. By taking sustainability into account we will move toward a future where the health of current, as well as future generations, is preserved.

ETHICAL CONSIDERATIONS

In addition to health benefits and increased sustainability, consuming a plant-based diet offers us a chance to reflect more objectively on the lives of the animals that have previously ended up on our plate. It is no coincidence that this section is the last in the book. What follows is entirely optional reading material, provided for those interested in exploring additional reasons for a plant-based diet beyond health and sustainability. Farming is not the same today as it once was. It is no longer common to see small family owned farms. Necessity has changed farming practices as the demand for agricultural products has increased. Arguably, the animals we farm today have been transformed as well. Physically, we have changed them to be more efficient producers of product. Mentally they remain much the same. The emerging body of research on the inner lives of farm animals and how this information intersects with current farming practices is discussed further in the upcoming sections.

Animal Cognition

The differences between humans and other animals have been at the forefront of science and philosophy for thousands of years. Around 400 BCE Plato playfully defined man as the only creature to walk on two legs and have no feathers [89]. In response to this Diogenes, a contemporary philosopher of Plato, brought a shaved chicken into Plato's lecture room, setting it free, he exclaimed to the class "Behold! I've brought you Plato's man." Plato amended his definition to include a creature that walks on two legs, is naked and has broad nails. Over the following centuries, many theories have attempted to pinpoint what distinguishes humans from the rest of life on Earth. However, upon closer examination there appear to be more similarities than differences. As famously said by Voltaire when speaking out against the vivisection of animals (in the preanesthetic era) "Has nature arranged for this animal to have all the machinery of feelings only in order for it not to have any at all?"

Animals are capable of many things once believed to be exclusively human; animals feel pain and pleasure but also share more complex similarities in the way they think, known as cognitive ability. These abilities include the formation of strong social bonds, language, self-awareness, and problem solving, to name only a few [90–93]. The many similarities between humans and other animals are the result of a shared and intertwining evolutionary past that is present in varying degrees with all life on Earth.

A common perception of cognition is that humans sit at the top of a pyramid of intellect with other creatures below. However, this narrow view does not take into account forms of cognition incomparable to our own. As paraphrased from *The Inner Life of Animals* a book by leading primatologist

> "Has nature arranged for this animal to have all the machinery of feelings only in order for it not to have any at all?"
> —Voltaire

Dr. Frans de Waal: You would not think of yourself as less intelligent than a squirrel simply because you are not able to remember the location of hundreds of buried nuts and would not presume your perception of surroundings better than that of an echolocating bat. Instead of thinking of cognition and intellect as a one-way path to the top, Dr. Frans de Waal would have us view it as a bush with many different branching possibilities [94]. While research examining the cognition of other animals has seen much development over the years, research on farm animals predominantly focuses on aspects related to their productivity. Contrary to popular belief, the domestication of farm animals has not left them mentally or behaviorally very different from their wild ancestors. In fact, what little research has been done on farm animals has shown they are capable of complex mental functions. Dr. Lori Marino, a neuroscientist and leading expert in animal behavior and intelligence, has complied several extensive reviews investigating cognition in farmed animals [91,92,95]. The abilities of most commonly farmed animals including chickens, pigs, and cows are discussed here.

> As paraphrased from The Inner Life of Animals a book by leading primatologist Dr. Frans de Waal: You would not think of yourself as less intelligent than a squirrel simply because you are not able to remember the location of hundreds of buried nuts.

Chickens: Chickens were domesticated from the red jungle fowl (*Gallus gallus*) native to India and southeast Asia. Like the red jungle fowl they descended from, chickens prefer to congregate in small groups. As one might expect from social animals, they have been shown to possess the ability to recognize and distinguish between familiar and unfamiliar chickens [96]. This type of recognition indicates a basis for forming social relationships typical for animals living in groups [91]. For example, if you were not able to recognize someone that you had a positive interaction with you would not be able to seek them out, or if they harmed you in the past, you would not be able to avoid them in the future. In addition to many anecdotal stories, research has shown chickens have distinct and often colorful personalities, exhibiting varying degrees of traits such as curiosity and boldness [91]. Chickens have strong maternal instincts and experience high levels of stress when their chicks are perceived to be in danger. Unlike people, chickens have the unique ability to simultaneously focus their eyes on items close by and far away [97]. They are also able to see additional colors in the spectrum, making the world they see likely very different than our own [97].

> Chickens have strong maternal instincts and experience high levels of stress when their chicks are perceived to be in danger.

Pigs: Pigs are highly social animals, who, like the wild boar they are descended from, also prefer to reside in small groups [98]. Like many other social animals, pigs enjoy playing with one another by engaging in play fighting or chasing games. This behavior, along with a propensity for playing with toys, is similar to play found in dogs [99]. Pigs have a remarkable sense of smell that is much more refined than humans. With their fine-tuned snouts, they are able to view the world through an olfactory lens, using this sense in many ways, including navigating social situations by sniffing out the moods of fellow pigs during tense encounters [100].

> Like many other social animals, pigs enjoy playing with one another by engaging in play fighting or chasing games. This behavior, along with a propensity for playing with toys, is similar to play found in dogs [99].

Cattle: Cattle are originally descended from the now extinct Auroch (*Bos primigenius*) [101]. The Auroch is an ancient species with a long relationship with people. While domestication dates back much further, evidence suggests aurochs played a role in roman culture over 2000 years ago where they were used during battle [101]. Aurochs existed in the wild up until the 1600s when they were hunted to extinction [101]. Little is known about the behavior and cognition of the Auroch, leaving cattle along with their other descendants to fill in the gaps. Cattle are social animals and although it is unclear what exactly a "natural" herd size may be, they are able to form the strongest bonds in smaller herds. This ability is often diminished once herds reach a larger size [102]. Smaller herds may offer a greater chance for interaction in much the same as living in a small town compared to a big city does. Cows have distinct personality traits such as gregarious, anxious, sociable, and adventurous. Within herds cows show a preference for the company of certain individuals and tend to avoid others [103]. Interestingly, they tend to seek out relationships with others of a similar disposition [104,105]. Relationships and company are particularly important in times of distress, as research has shown both cows and bulls are calmed by the presence of others [106]. Cows have even been found to bond with animals outside of their own species such as sheep and dogs [107–109]. Cows also have the capability to form bonds with people, although given the nature of most interactions, bonding in this context can be difficult. When relationships are positive, cows enjoy being interactions such as being scratched behind the ears, seeking out this attention similar to how a cat or a dog might [110].

> Relationships and company are particularly important in times of distress, as research has shown both cows and bulls are calmed by the presence of others.

Undeniably the human mind is complex and has accomplished many amazing things both good and bad. This section aimed to highlight some of the many attributes we share with other animals and some of the curious and intriguing traits we do not. As Charles Darwin once said "The difference in mind between man and the higher animals, great as it is, certainly is one of degree and not of kind." Due to the intellect we possess we are able to contemplate the thoughts and feeling of other species. For those lucky enough to have access to a reliable and affordable food supply we are put in the unique position to choose what we eat according to a set of ethical/moral values and not purely out of necessity.

> "The difference in mind between man and the higher animals, great as it is, certainly is one of degree and not of kind."
> —Charles Darwin

The Current State of Livestock Production

Industrialized Farming

Industrialized farming, often referred to as conventional or factory farming, is defined by a high concentration of animals per unit of space. Food to supply the animals on these farms is prominently grown at a separate location and transported to animals kept in confined conditions called feedlots. Sometimes on industrialized farms these feedlots are used in combination with pastures and outdoor runs. However, for many animals in this system, particularly chickens, outdoor access is extremely rare. Industrialized farming systems are becoming increasingly popular for rearing animals. According to a report by the Food and Agriculture Organization around 20 years ago, industrialized farming contributed to 72% of the global poultry and 55% of the pork meat, around 66% of global eggs supply

> Globally, the population of cows living in industrial farms outnumber the human population of Canada.

20 year change in industrial animal production

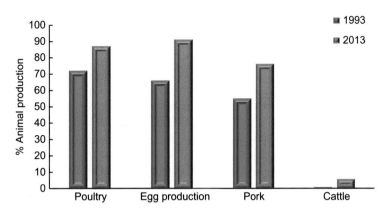

FIG. 6.2   Change in the percent of animals produced using industrial production over 20 years.

and negligible amounts of cattle [111]. As of 2013, 87% of poultry, 91% of chicken eggs, 76% of pork, and 6% of large ruminant meat production such as cattle are now housed in industrial farm facilities [112] (Fig. 6.2). In regard to large ruminants such as cattle, 6% may still appear to be a small amount, however, to put this number into perspective, 6% of global cattle production outnumbers the human population of Canada. Having such a high concentration of animals living in close quarters, as is the case in factory farms, typically reduces quality of life and can also result in the spread of disease, see Box 6.5 for more information.

BOX 6.5 Disease and Environmental Impact from Industrial Operations

Industrial farms contribute to a host of environmental and health issues. The often large separation between where feed is grown and where cattle are raised mean that manure generated can greatly surpass the need for fertilizer in a region [113]. To accommodate this overflow, manure is often stored in what is referred to as "lagoons" which are essentially large, open pits. These manure lagoons are prone to spills which are extremely damaging to lake and river ecosystems, killing off many species that live within them [113]. Diseases can also be spread through fecal matter making spills a public health concern [114].

Both antibiotic-resistant bacterial infections and viral infections such as swine and bird flu can be spread to humans through infected animal feces as well as through improperly cooked and handled meats [114,115]. Many antibiotics that were once able to treat most bacterial infections are becoming increasingly ineffective [114]. These bacterial infections that are antibiotic-resistant are called "superbugs." Super bugs were created in places like the agricultural industry where animals are fed a steady stream of antibiotics. In industrialized food systems antibiotics are given not only to sick animals but also to healthy ones to improve growth [116]. This large scale use has given bacteria, like salmonella, the opportunity to evolve antibiotic resistance [114]. Viral infections can be extremely dangerous, with bird flu causing death in over 50% of human cases [115]. In the same way that an infection will spread more quickly in a highly populated area like a city, disease will spread more rapidly when animals are kept in high density conditions, as in industrial systems.

Conditions in Industrial Settings

The fast-paced growth of industrial farming is particularly troubling given the conditions that animals within the system face. It is a common occurrence for chickens in industrial farms to spend the entirety their lives indoors. For meat chickens, which have been bred to gain body mass quickly, the time from hatching to slaughter is approximately 41 days [117]. This fast growth rate can result in their bones being unable keep up with the accelerated pace of development, causing them to break under the weight of their bodies [117]. Due to the high density at which chickens are kept, injuries often go unnoticed, leaving them vulnerable to painful infections [118]. The extraordinary sensory abilities and eccentricities of chickens are stifled in the cramped, typically indoor environments found in industrialized systems (so-called "battery hens"). Boredom and social tensions in commercial settings cause chickens to engage in self-mutilation, such as feather plucking and cannibalism [119].

> The tip of a chickens' beak is used in a similar manor to fingers in that they are not only used to pick things up but are extremely sensitive to touch and can discriminate different textures.

Pigs in industrialized farms are often housed indoors with no outdoor access in tightly packed conditions, making exploring the complex world of smell they might otherwise engage in difficult. Ammonia and other gases given off by feces and urine are can be so strong they burn lesions in the lungs increasing susceptibility to pneumonia [113]. In more naturalized settings such as outdoor pens, sows typically make nests for their piglets out of straw and other insulating materials [120]. In all but a few American states and European countries it is considered standard practice in industrial farming to crate sows, forcing them to raise their young confined to an area not big enough to turn around or stand up [120]. In industrialized systems, for hygienic purposes, uncrated animals typically reside in areas with bare concrete or slotted floors, which can be painful for their cloven feet resulting in deformity [113].

> In more naturalized settings such as outdoor pens, sows typically make nests for their piglets out of straw and other insulating materials.

Rearing cattle in industrialized systems often involves some form of closed quarter confinement. In the U.S. this has often resulted in stall-style confinement where movement and normal social interaction with other cows is prevented [121]. Keeping animals in stalls and restricting them from pasture means that food must first be harvested, then fed to cattle. These diets are often high in grain, a food that the digestive system of the cow is not well adapted to consuming, and often results in gastrointestinal distress such as diarrhea, acidosis, and other serious and uncomfortable disorders [122]. Like pigs and chickens, cattle often spend the majority of their lives on hard concrete floors, very different from the soft grassy pastures they have evolved to stand in [121]. Although this is done for hygienic purposes, living on a hard floor can lead to serious health problems such as lameness and foot lesions resulting in pain and infection [121].

Selective breeding of dairy cows for milk production along with unsanitary conditions can result in mastitis, a painful condition caused by blocked milk flow [121].

> Cows form strong maternal bonds with their offspring and when separated will call relentlessly to them.

A little discussed aspect of the dairy industry is its role in supplying animals to the veal industry. In order for cows to keep producing milk, like any mammal, they must continue to have offspring. Male calves cannot produce milk and are not economically viable to raise for beef. This economic catch typically leaves them fated for the veal industry [123]. Veal is largely viewed as the poster child of cruel animal treatment. In order to prevent meat from becoming tough and red, calves are often kept in confinement where there is little to no room to move and gain tough muscle [124]. They spend their lives inside, unable to exhibit natural behavior, play or socialize. In some cases even eye contact is prevented by the solid walls of the crate [124]. Removing calves from the care of their mothers can be very traumatic for both individuals, especially when done before natural weening takes place. Cows form strong maternal bonds with their offspring and when separated will call relentlessly for them. They will also show signs of intense physical distress such as excessive urination and diarrhea as well as repetitive behaviors like head shaking [125]. Calves show similar signs of distress when removed from their mothers. In studies where calves were raised to adulthood, those that were not suckled became more anxious adults and had poorer social integration with their peers [126]. Interestingly, this is comparable to human studies where children who were breastfed were less likely to experience mental health issues later in life [127]. As the result of awareness campaigns, many veal farming institutions have greatly improved their welfare standards, particularly in regions where there is prohibition of veal crates backed by legislation and consumer support. However, there is still much that must be done in ensure humane treatment of all of the animals. Avoiding dairy, veal, chicken, and pork products is one way that you can make a difference.

Drivers of Industrial Farming

So why is industrial farming increasingly becoming the norm when it comes to raising animals? While many factors are certainly at work, the demand for cheaper products is the main driving force. This demand has put farmers in a difficult position where they must decide between converting to industrialized farming or not having competitive prices. In areas where industrial farms are built, the relative price of a product is driven down and subsequently the quantity of product required to make a living is increased. This situation has edged out a lot of small scale farmers who are not able to industrialize their practices and compete in the market [128]. In countries where small scale farmers are responsible for the majority of food production, the increase of industrial systems has greatly disadvantaged many of these people [129].

In many countries many small scale farmers have been disadvantaged by the increase of industrial systems.

Nonindustrial Farms

Different farms have varying degrees of animal welfare and it would be untrue to say that all farms uphold the standards discussed in the above sections industrial or not. Many farmers care very much for the animals they raise and try to make their lives as comfortable as possible. Different farming methods such as organic and biodynamic approaches are required to uphold a higher standard of welfare for animals. These standards include animal density restrictions and access to the outdoors [130,131]. However, it is worth noting that even under the best conditions, where animals can live a relatively wild existence they still typically live a fraction of their natural lifespan and it is questionable whether a humane method of killing truly exists.

Parting Words

Individual food choice has the power to create a ripple effect into the food system through shifting demand. The wellbeing and survival of many species depend on what our individual choices are. While government, food companies, and other individuals in the food system have a responsibility to produce change, decisions made at the level of the individual can be the most powerful determinants of change. Capitalism may have many shortfalls but one benefit is the ability to produce change through demand. This ability can be harnessed to incentivize the production of healthier, more environmentally conscious and ethically produced foods. This chapter is not intended to force a set of moral judgments onto people, we have placed it here simply to provide one viewpoint. While not every or possibly any environmental aspect or ethical reasoning discussed here will connect with every individual, perhaps in instances where it does, this will add an additional incentive to maintain the healthy diet and lifestyle choice that is the Portfolio diet.

REFERENCES

[1] Bernstein AM, Sun Q, Hu FB, Stampfer MJ, Manson JE, Willett WC. Coronary heart disease major dietary protein sources and risk of coronary heart disease in women. Circulation 2010;122(9):876–83. https://doi.org/10.1161/CIRCULATIONAHA.109.915165.

[2] Pan A, Sun Q, Bernstein AM, Manson JE, Willett WC, Hu FB. Changes in red meat consumption and subsequent risk of type 2 diabetes mellitus. JAMA Intern Med 2013;173(14):1328. https://doi.org/10.1001/jamainternmed.2013.6633.

[3] Sun Q, Sun Q, Bernstein AM, Schulze MB, Manson JE, Stampfer MJ, et al. Red meat consumption and mortality. Arch Intern Med 2012;172(7):555. https://doi.org/10.1001/archinternmed.2011.2287.

[4] Clarys P, Deliens T, Huybrechts I, Deriemaeker P, Vanaelst B, Keyzer W, et al. Comparison of nutritional quality of the vegan, vegetarian, semi-vegetarian, pesco-vegetarian and omnivorous diet. Nutrients 2014;6(3):1318–32. https://doi.org/10.3390/nu6031318.

[5] Tilman D, Clark M. Global diets link environmental sustainability and human health. Nature 2014;515(7528):518–22. https://doi.org/10.1038/nature13959.

[6] Daniel CR, Cross AJ, Koebnick C, Sinha R. Trends in meat consumption in the USA. Public Health Nutr 2011;14(4):575–83. https://doi.org/10.1017/S1368980010002077.

[7] Craig WJ. Nutrition concerns and health effects of vegetarian diets. Nutr Clin Pract 2010;25(6):613–20. https://doi.org/10.1177/0884533610385707.

[8] Seldin MM, Meng Y, Qi H, Zhu W, Wang Z, Hazen SL, et al. Trimethylamine N-oxide promotes vascular inflammation through signaling of mitogen-activated protein kinase and nuclear factor-κB. J Am Heart Assoc 2016;5(2):e002767. https://doi.org/10.1161/JAHA.115.002767.

[9] Tang WHW, Wang Z, Levison BS, Koeth RA, Britt EB, Fu X, et al. Intestinal microbial metabolism of phosphatidylcholine and cardiovascular risk. N Engl J Med 2013;368(17):1575–84. https://doi.org/10.1056/NEJMoa1109400.

[10] Oellgaard J, Winther SA, Hansen TS, Rossing P, von Scholten BJ. Trimethylamine N-oxide (TMAO) as a new potential therapeutic target for insulin resistance and cancer. Curr Pharm Des 2017;23(25):3699–712. https://doi.org/10.2174/1381612823666170622095324.

[11] Dambrova M, Latkovskis G, Kuka J, Strele I, Konrade I, Grinberga S, et al. Diabetes is associated with higher trimethylamine N-oxide plasma levels. Exp Clin Endocrinol Diabetes 2016;124(4):251–6. https://doi.org/10.1055/s-0035-1569330.

[12] Farvid MS, Cho E, Chen WY, Eliassen AH, Willett WC. Adolescent meat intake and breast cancer risk. Int J Cancer 2015;136(8):1909–20. https://doi.org/10.1002/ijc.29218.

[13] Bouvard V, Loomis D, Guyton KZ, Grosse Y, Ghissassi F, Benbrahim-Tallaa L, Guha N, Mattock H, Straif K. Agency for Research on Cancer Monograph Working Group I. Carcinogenicity of consumption of red and processed meat. Lancet Oncol 2015;16:1599–600.

[14] World Health Organization. A healthy lifestyle. Available at, http://www.euro.who.int/en/health-topics/disease-prevention/nutrition/a-healthy-lifestyle; 2018. (Accessed March 20, 2018).

[15] World Health Oraganization. Diet. Available at, http://www.who.int/dietphysicalactivity/diet/en/; 2016. (Accessed January 1, 2016).

[16] The European Food Information Council. Food-based dietary guidelines in Europe. Available at, http://www.eufic.org/en/healthy-living/article/food-based-dietary-guidelines-in-europe; 2009. (Accessed March 20, 2018).

[17] American Diabetes Association®. Protein Foods: American Diabetes Association®. Available at, http://www.diabetes.org/food-and-fitness/food/what-can-i-eat/making-healthy-food-choices/meat-and-plant-based-protein.html; 2017. (Accessed March 20, 2018).

[18] Canadian Diabetes Association. Basic meal planning | Diabetes Canada. Available at, http://www.diabetes.ca/diabetes-and-you/healthy-living-resources/diet-nutrition/basic-meal-planning; 2017. (Accessed March 20, 2018).

[19] Canadian Cardiovascular Society-Medical. Nutritional requirements for adults: nutrition tools by CCS medical. Available at, https://ccsmed.com/living-healthy/article/dietary-recommendations-for-adults/; 2017. (Accessed March 20, 2018).

[20] HEART UK. Low cholesterol diet | High cholesterol foods | HEART UK | Expert advice from HEART UK. June 15, 2017, Retrieved from, https://heartuk.org.uk/cholesterol-and-diet/low-cholesterol-diets-and-foods; 2017. (Accessed March 20, 2018).

[21] Catapano AL, et al. 2016 ESC/EAS guidelines for the management of dyslipidaemias. Rev Esp Cardiol (Engl Ed) 2017;70:115.

[22] World Health Organization. IARC Monographs evaluate consumption of red meat and processed meat and cancer risk. Int Agency Res Cancer; 2015. p. 1–2. http://www.who.int/features/qa/cancer-red-meat/en/.

[23] American Heart Association. What is a healthy diet? Recommended serving infographic. Available at, https://healthyfor-good.heart.org/eat-smart/infographics/what-is-a-healthy-diet-recommended-serving-infographic; 2017. (Accessed March 20, 2018).

[24] The Offical Dietary Guidlleines for Americans. 2015-2020 Dietary Guidelines for Americans. Retrieved from, http://health.gov/dietaryguidelines/2015/guidelines/; 2015.

[25] Government of Canada. Canada's Food Guide. Retrieved from https://food-guide.canada.ca/en/; 2019.

[26] The Official Dutch Dietary Guidelines. Richtlijnen Schijf van Vijf. Retrieved from, http://www.voedingscentrum.nl/Assets/Uploads/voedingscentrum/Documents/Consumenten/SchijfvanVijf2016/VC_Richtlijnen_Schijf_van_Vijf_2016.pdf; 2015.

[27] Belgium Dietary Guidelines. Gezonde voeding: de praktische gidsen. Retrieved from, http://www.fao.org/nutrition/education/food-dietary-guidelines/regions/countries/belgium/en/; 2017.

[28] Chinese Dietary Guidelines. Core Recommendations—Dietary Guidelines for Chinese Residents. Retrieved from, http://dg.cnsoc.org/article/04/8a2389fd5520b4f30155be01beb82724.html; 2016.

[29] European Food Information Council. Food-Based Dietary Guidelines in Europe: (EUFIC). June 5, 2017, Retrieved from, http://www.eufic.org/en/healthy-living/article/food-based-dietary-guidelines-in-europe; 2009.

[30] Hooper L, Thompson RL, Harrison RA, Summerbell CD, Ness AR, Moore HJ, et al. Risks and benefits of omega 3 fats for mortality, cardiovascular disease, and cancer: systematic review. BMJ 2006;332(7544):752–60. https://doi.org/10.1136/bmj.38755.366331.2F.

[31] Rizos EC, Elisaf MS. Does supplementation with omega-3 PUFAs add to the prevention of cardiovascular disease? Curr Cardiol Rep 2017;19(6):47. https://doi.org/10.1007/s11886-017-0856-8.

[32] Hu FB, Bronner L, Willett WC, Stampfer MJ, Rexrode KM, Albert CM, et al. Fish and omega-3 fatty acid intake and risk of coronary heart disease in women. JAMA 2002;287(14):1815–21. Retrieved from, http://www.ncbi.nlm.nih.gov/pubmed/11939867.

[33] Mozaffarian D, Rimm EB. Fish intake, contaminants, and human health. JAMA 2006;296(15):1885. https://doi.org/10.1001/jama.296.15.1885.

[34] Ascherio A, Rimm EB, Stampfer MJ, Giovannucci EL, Willett WC. Dietary intake of marine n-3 fatty acids, fish intake, and the risk of coronary disease among men. N Engl J Med 1995;332(15):977–83. https://doi.org/10.1056/NEJM199504133321501.

[35] Iso H, Rexrode KM, Stampfer MJ, Manson JE, Colditz GA, Speizer FE, et al. Intake of fish and omega-3 fatty acids and risk of stroke in women. JAMA 2001;285(3):304–12. Retrieved from, http://www.ncbi.nlm.nih.gov/pubmed/11176840.

[36] Morris MC, Manson JE, Rosner B, Buring JE, Willett WC, Hennekens CH. Fish consumption and cardiovascular disease in the physicians' health study: a prospective study. Am J Epidemiol 1995;142(2):166–75. Retrieved from, http://www.ncbi.nlm.nih.gov/pubmed/7598116.

[37] Domingo JL. Nutrients and chemical pollutants in fish and shellfish. Balancing health benefits and risks of regular fish consumption. Crit Rev Food Sci Nutr 2016;56(6):979–88. https://doi.org/10.1080/10408398.2012.742985.

[38] Burr ML, Ashfield-Watt PAL, Dunstan FDJ, Fehily AM, Breay P, Ashton T, et al. Lack of benefit of dietary advice to men with angina: results of a controlled trial. Eur J Clin Nutr 2003;57(2):193–200. https://doi.org/10.1038/sj.ejcn.1601539.

[39] Michaëlsson K, Wolk A, Langenskiöld S, Basu S, Warensjö Lemming E, Melhus H, Byberg L. Milk intake and risk of mortality and fractures in women and men: cohort studies. BMJ (Clinical Research Ed) 2014;349:g6015. https://doi.org/10.1136/BMJ.G6015.

[40] Cui X, Wang L, Zuo P, Han Z, Fang Z, Li W, Liu J. D-Galactose-caused life shortening in Drosophila melanogaster and Musca domestica is associated with oxidative stress. Biogerontology 2004;5(5):317–26. https://doi.org/10.1007/s10522-004-2570-3.

[41] Cui X, Zuo P, Zhang Q, Li X, Hu Y, Long J, et al. Chronic systemic D-galactose exposure induces memory loss, neurodegeneration, and oxidative damage in mice: protective effects of R-α-lipoic acid. J Neurosci Res 2006;83(8):1584–90. https://doi.org/10.1002/jnr.20845.

[42] Feskanich D, Willett WC, Stampfer MJ, Colditz GA. Milk, dietary calcium, and bone fractures in women: a 12-year prospective study. Am J Public Health 1997;87(6):992–7. Retrieved from, http://www.ncbi.nlm.nih.gov/pubmed/9224182.

[43] Mullie P, Pizot C, Autier P. Daily milk consumption and all-cause mortality, coronary heart disease and stroke: a systematic review and meta-analysis of observational cohort studies. BMC Public Health 2016;16(1):1236. https://doi.org/10.1186/s12889-016-3889-9.

[44] Malekinejad H, Rezabakhsh A. Hormones in dairy foods and their impact on public health—a narrative review article. Iran J Public Health 2015;44(6):742–58. Retrieved from, http://www.ncbi.nlm.nih.gov/pubmed/26258087.

[45] Voskuil DW, Vrieling A, van't Veer LJ, Kampman E, Rookus MA. The insulin-like growth factor system in cancer prevention: potential of dietary intervention strategies. Cancer Epidemiol Biomark Prev 2005;14(1):195–203. Retrieved from, http://www.ncbi.nlm.nih.gov/pubmed/16041870.

[46] Chan JM, Stampfer MJ, Giovannucci E, Gann PH, Ma J, Wilkinson P, et al. Plasma insulin-like growth factor-I and prostate cancer risk: a prospective study. Science (New York, NY) 1998;279(5350):563–6. Retrieved from, http://www.ncbi.nlm.nih.gov/pubmed/9438850.

[47] Hsing AW, Tsao L, Devesa SS. International trends and patterns of prostate cancer incidence and mortality. Int J Cancer 2000;85(1):60–7. Retrieved from, http://www.ncbi.nlm.nih.gov/pubmed/10585584.

[48] Aune D, Lau R, Chan DSM, Vieira R, Greenwood DC, Kampman E, Norat T. Dairy products and colorectal cancer risk: a systematic review and meta-analysis of cohort studies. Ann Oncol 2012;23(1):37–45. https://doi.org/10.1093/annonc/mdr269.

[49] David Spence J. Dietary cholesterol and egg yolk should be avoided by patients at risk of vascular disease. J Transl Intern Med 2016;4(1):20–4. https://doi.org/10.1515/jtim-2016-0005.

[50] Fernandez ML. Dietary cholesterol provided by eggs and plasma lipoproteins in healthy populations. Curr Opin Clin Nutr Metab Care 2006;9(1):8–12. https://doi.org/10.1097/01.mco.0000171152.51034.bf.

[51] Howell WH, McNamara DJ, Tosca MA, Smith BT, Gaines JA. Plasma lipid and lipoprotein responses to dietary fat and cholesterol: a meta-analysis. Am J Clin Nutr 1997;65(6):1747–64. https://doi.org/10.1093/ajcn/65.6.1747.

[52] Hu FB, Stampfer MJ, Rimm EB, Manson JE, Ascherio A, Colditz GA, et al. A prospective study of egg consumption and risk of cardiovascular disease in men and women. JAMA 1999;281(15):1387–94.

[53] Weggemans RM, Zock PL, Tai ES, Ordovas JM, Molhuizen HOF, Katan MB. ATP binding cassette G5 C1950G polymorphism may affect blood cholesterol concentrations in humans. Clin Genet 2002;62(3):226–9.

[54] Shin JY, Xun P, Nakamura Y, He K. Egg consumption in relation to risk of cardiovascular disease and diabetes: a systematic review and meta-analysis. Am J Clin Nutr 2013;98(1):146–59. https://doi.org/10.3945/ajcn.112.051318.

[55] Zhong VW, Van Horn L, Allen NB. Dietary cholesterol or egg consumption and cardiovascular outcomes—reply. JAMA 2019;322(5):467. https://jamanetwork.com/journals/jama/fullarticle/2740714.

[56] World Health Organization. COP24 Special Report: Special Report Health and Climate. In: WHO. World Health Organization; 2018.

[57] Willett W, Rockström J, Loken B, Springmann M, Lang T, Vermeulen S, et al. Food in the Anthropocene: the EAT–*Lancet* commission on healthy diets from sustainable food systems. Lancet. https://doi.org/10.1016/S0140-6736(18)31788-4; 2019.

[58] Watson SB, Miller C, Arhonditsis G, Boyer GL, Carmichael W, Charlton MN, et al. The re-eutrophication of Lake Erie: harmful algal blooms and hypoxia. 2016. https://doi.org/10.1016/j.hal.2016.04.010.

[59] Mekonnen MM, Hoekstra AY. Value of water research report series no. 48 volume 1: main report value of water. Retrieved from, http://waterfootprint.org/media/downloads/Report-48-WaterFootprint-AnimalProducts-Vol1_1.pdf; 2010.

[60] FAO. Livestocks long shadow: environmental issues and options. Retrieved from, ftp://ftp.fao.org/docrep/fao/010/a0701e/a0701e.pdf; 2006.

[61] FAO. Key facts and findings. Retrieved from, http://www.fao.org/news/story/en/item/197623/icode/; 2017.

[62] IPCC. Summary for policymakers. In: Global Warming of 1.5°C. An IPCC special report on the impacts of global warming of 1.5°C above pre-industrial levels and related global greenhouse gas emission pathways, in the context of strengthening the global response to the threat of climate change, sustainable development, and efforts to eradicate poverty. IPCC SR15. Retrieved from http://www.ipcc.ch/report/sr15/; 2018.

[63] IUCN. Red list of Mediterranean endemic freshwater fish. 20081–2. Retrieved from, papers2://publication/uuid/B091CA46-477E-4C3E-BE91-277E9094DD5E.

[64] FAO. Agriculture, forestry and other land use emissions by sources and removals by sinks. Retrieved from, http://www.fao.org/docrep/019/i3671e/i3671e.pdf; 2014.

[65] Tlman D, Clark M. Global diets link environmental sustainability and human health. Nature 2014;515(7528):518–22.

[66] The IPPC. Climate change 2014 mitigation of climate change working group III contribution to the fifth assessment report of the intergovernmental panel on climate change. Retrieved from, https://www.ipcc.ch/report/ar5/wg3/; 2014.

[67] European Commission. Institute for prospective technological studies european science and technology observatory. Retrieved from, http://ec.europa.eu/environment/ipp/pdf/eipro_report.pdf; 2006.

[68] Garnett T, Godde C, Muller A, Röös E, Smith P, De Boer I, et al. Grazed and confused-ruminating on cattle, grazing systems, methane, nitrous oxide, the soil carbon sequestration question—and what it all means for greenhouse gas emission. Food Climate Research Network. Retrieved from, https://www.fcrn.org.uk/sites/default/files/project-files/fcrn_gnc_report.pdf; 2017.

[69] Capper JL. Is the grass always greener? Comparing the environmental impact of conventional, natural and grass-fed beef production systems. Animals 2012;2(4):127–43. https://doi.org/10.3390/ani2020127.

[70] Garnett T. Where are the best opportunities for reducing greenhouse gas emissions in the food system (including the food chain)? Food Policy 2011;36:S23–32. https://doi.org/10.1016/j.foodpol.2010.10.010.

[71] Hallstrom E, Carlsson-Kanyama A, Borjesson P. Environmental impact of dietary change: a systematic review. J Clean Prod 2015;91:1–11. https://doi.org/10.1016/j.jclepro.2014.12.008.

[72] FAO. Cattle ranching and deforestation. In: The Livestock Policy Brief, vol .3; 2005. p. 1–8. Retrieved from, http://www.fao.org/3/a-a0262e.pdf.

[73] FAO. Livestock and landscapes. Retrieved from, http://www.fao.org/docrep/018/ar591e/ar591e.pdf; 2014.

[74] Singer P, Mason J. The way we eat: why our food choices matter. Anim Liberation Philos Policy J 2006;IV(1):1–8. https://doi.org/9781605296074.

[75] Venter O, Brodeur NN, Nemiroff L, Belland B, Dolinsek IJ, Grant JWA. Threats to Endangered species in Canada. BioScience 2006;56(11):903–10. https://doi.org/10.1641/0006-3568(2006)56[903:ttesic]2.0.co;2.

[76] Federal Provincial and Territorial Governments Canada. Canadian Biodiversity: Ecosystem Status and Trends 2010. Federal, Provincial, and Territorial Governments of Canada; 2010. https://doi.org/vi+142p.

[77] Benis K, Ferr P. Potential mitigation of the environmental impacts of food systems through urban and peri-urban agriculture (UPA)—a life cycle assessment approach. 2017. https://doi.org/10.1016/j.jclepro.2016.05.176.

[78] FAO Fisheries and Aquculture Department. The state of world fisheries and aquaculture. Aquaculture 2006;162, https://doi.org/978-92-5-105568-7.

[79] Stachowicz JJ, Whitlatch EB, Osman N, Duffy JE, Folke C, Halpern BS, et al. Species diversity and invasion resistance in a marine ecosystem. Science 1999;286(5444):1577–9. https://doi.org/10.1126/science.286.5444.1577.

[80] Schrank WE. The Newfoundland fishery: ten years after the moratorium. Mar Policy 2005;29(5):407–20. https://doi.org/10.1016/J.MARPOL.2004.06.005.

[81] Alder J, Sumaila UR. Western Africa: a fish basket of Europe past and present. J Environ Dev 2004;13(2):156–78. https://doi.org/10.1177/1070496504266092.

[82] Holmer M. Environmental issues of fish farming in offshore waters: perspectives, concerns and research needs. Aquac Environ Interact 2010;1(1):57–70. https://doi.org/10.3354/aei00007.

[83] Naylor RL, Goldburg RJ, Primavera JH, Kautsky N, Beveridge MCM, Clay J, et al. Effect of aquaculture on world fish supplies. Nature 2000;405(6790):1017–24. https://doi.org/10.1038/35016500.

[84] Sonesson U, Davis J, Ziegler F. Food production and emissions of greenhouse gases: an overview of the climate impact of different product groups. SIK Report. No 802. The Swedish Institute for Food and Biotechnology. Retrieved from, http://www.diva-portal.org/smash/get/diva2:943607/FULLTEXT01.pdf; 2010.

[85] FAO. FAO-statistics. June 12, 2018, Retrieved from, http://www.fao.org/faostat/en/#data/QC/visualize; 2016.

[86] Smetana S, Mathys A, Knoch A, Heinz V. Meat alternatives: life cycle assessment of most known meat substitutes. Int J Life Cycle Assess 2015;20(9):1254–67. https://doi.org/10.1007/s11367-015-0931-6.

[87] Blonk H, Kool A, Luske B, Waart SD. Environmental effects of protein-rich food products in the Netherlands Consequences of animal protein substitutes. Gouda: Blonk Milieu Advies; 2008. p. 1–19. December, Retrieved from, http://www.blonkconsultants.nl/wp-content/uploads/2016/06/english-summary-protein-rich-products.pdf.

[88] FAO. Key facts on food loss and waste you should know! I SAVE FOOD: Global Initiative on Food Loss and Waste Reduction I Food and Agriculture Organization of the United Nations. June 20, 2018, Retrieved from, http://www.fao.org/save-food/resources/keyfindings/en/; 2018.

[89] Hicks RD. Lives of eminent philosophers part II. Harvard University Press; 1925. Retrieved from, https://ryanfb.github.io/loebolus-data/L185.pdf.

[90] de Waal FBM. The antiquity of empathy. Science (New York, NY) 2012;336(6083):874–6. https://doi.org/10.1126/science.1220999.

[91] Marino L. Thinking chickens: a review of cognition, emotion, and behavior in the domestic chicken. Anim Cogn 2017;20(2):127–47. https://doi.org/10.1007/s10071-016-1064-4.

[92] Marino L, Colvin CM. Thinking pigs: a comparative review of cognition, emotion, and personality in *Sus domesticus*. Int J Comp Psychol 2015;28(1). Retrieved from, https://escholarship.org/uc/item/8sx4s79c.

[93] Pollick AS, de Waal FBM. Ape gestures and language evolution. Proc Natl Acad Sci USA 2007;104(19):8184–9. https://doi.org/10.1073/pnas.0702624104.

[94] de Waal F. Are we smart enough to know how smart other animals are. 1st ed. New York: W.W Norton and Company; 2016.

[95] Marino L, Allen K. The psychology of cows. Anim Behav Cogn 2017;4(44):474–98. https://doi.org/10.26451/abc.04.04.06.2017.

[96] D'Eath R, Stone R. Chickens use visual cues in social discrimination: an experiment with coloured lighting. Appl Anim Behav Sci 1999;62(2–3):233–42. https://doi.org/10.1016/S0168-1591(98)00216-0.

[97] Stamp Dawkins M, Woodington A. Distance and the presentation of visual stimuli to birds. Anim Behav 1997;54:1019–25. Retrieved from, http://users.ox.ac.uk/~snikwad/resources/Distance.pdf.

[98] Maselli V, Rippa D, Russo G, Ligrone R, Soppelsa O, D'Aniello B, et al. Wild boars' social structure in the Mediterranean habitat. Ital J Zool 2014;81(4):610–7. https://doi.org/10.1080/11250003.2014.953220.

[99] Horback K. Nosing around: play in pigs. Anim Behav Cogn 2014;2014(12):186–96. https://doi.org/10.12966/abc.

[100] McGlone JJ. Olfactory signals that modulate pig aggressive and submissive behavior. 1990. p. 86–109. Retrieved from, https://www.cabdirect.org/cabdirect/abstract/19902215196.

[101] Cis van Vuure S. Retracting the aurochs history, morphology and ecology of an extinct wild ox. Q Rev Biol 2006;(3)81. https://doi.org/10.1086/509405.

[102] Kondo S, Sekine J, Okubo M, Asahida Y. The effect of group size and space allowance on the agonistic and spacing behavior of cattle. Appl Anim Behav Sci 1989;24(2):127–35. https://doi.org/10.1016/0168-1591(89)90040-3.

[103] Gygax L, Neisen G, Wechsler B. Socio-spatial relationships in dairy cows. Ethology 2010;116(1):10–23. https://doi.org/10.1111/j.1439-0310.2009.01708.x.

[104] Boyland NK, Mlynski DT, James R, Brent LJN, Croft DP. The social network structure of a dynamic group of dairy cows: from individual to group level patterns. Appl Anim Behav Sci 2016;174:1–10. https://doi.org/10.1016/j.applanim.2015.11.016.

[105] Müller R, Schrader L, Miuilerl R. Behavioural consistency during social separation and personality in dairy cows. Behaviour 2005;14210(9):1289–306. Retrieved from, http://www.jstor.org/stable/4536301.

[106] Kikusui T, Winslow JT, Mori Y. Social buffering: relief from stress and anxiety. Philos Trans R Soc Lond Ser B Biol Sci 2006;361(1476):2215–28. https://doi.org/10.1098/rstb.2006.1941.

[107] Anderson DM, Hulet CV, Shupe WL, Smith JN, Murray LW. Response of bonded and non-bonded sheep to the approach of a trained border collie. Appl Anim Behav Sci 1988;21(3):251–7. https://doi.org/10.1016/0168-1591(88)90114-1.

[108] Coppinger RP, Smith CK, Miller L. Observations on why mongrels may make effective livestock protecting dogs. J Range Manag 2006;38(6):560–1. https://doi.org/10.2307/3899754.

[109] Petherick JC, Doogan VJ, Venus BK, Holroyd RG, Olsson P. Quality of handling and holding yard environment, and beef cattle temperament: 2. Consequences for stress and productivity. Appl Anim Behav Sci 2009;120:28–38. https://doi.org/10.1016/j.applanim.2009.05.009.

[110] Moran J. Calf rearing: a guide to rearing calves in Australia. Dept. of Agriculture; 1993. Retrieved from, http://agris.fao.org/agris-search/search.do?recordID=AU9430106.

[111] FAO. World Livestock Production Systems FAO ANIMAL Carlos Seré and Henning Steinfeld in collaboration with. Retrieved from, http://www.fao.org/3/a-w0027e.pdf; 1996.

[112] Herrero M, Havlík P, Valin H, Notenbaert A, Rufino MC, Thornton PK, et al. Biomass use, production, feed efficiencies, and greenhouse gas emissions from global livestock systems. Proc Natl Acad Sci USA 2013;110(52):20888–93. https://doi.org/10.1073/pnas.1308149110.

[113] Horrigan L, Lawrence RS, Walker P. How sustainable agriculture can adress the environmental and human health harms of industrial agriculture. Environ Health Perspect 2002;110(5):445–56. https://doi.org/10.1289/ehp.02110445.

[114] Centers for Disease Control and Prevention. Antibiotic resistance | NARMS | CDC. June 21, 2018, Retrieved from, https://www.cdc.gov/narms/faq.html; 2016.

[115] Centers for Disease Control and Prevention. Information on avian influenza | Avian influenza (flu). June 21, 2018, Retrieved from, http://www.cdc.gov/flu/avianflu/index.htm; 2014.

[116] Kemper N. Veterinary antibiotics in the aquatic and terrestrial environment. Ecol Indic 2008;8(1):1–13. https://doi.org/10.1016/J.ECOLIND.2007.06.002.

[117] Mignon-Grasteau S, Chantry-Darmon C, Boscher M-Y, Sellier N, Chabault-Dhuit M, Le Bihan-Duval E, Narcy A. Genetic determinism of bone and mineral metabolism in meat-type chickens: a QTL mapping study. Bone Rep 2016;5:43–50. https://doi.org/10.1016/j.bonr.2016.02.004.

[118] Toronto Vegetarian Association. Profile on chickens. May 16, 2018, Retrieved from, http://veg.ca/animal-issues/farm-animals/chickens-treatment/; 2018.

[119] Cheng H. Morphopathological changes and pain in beak trimmed laying hens. Worlds Poult Sci J 2006;62(1):41–52. https://doi.org/10.1079/WPS200583.

[120] Yun J, Valros A. Benefits of prepartum nest-building behaviour on parturition and lactation in sows—a review. Asian Australas J Anim Sci 2015;28(11):1519–24. https://doi.org/10.5713/ajas.15.0174.

[121] The Humane Society of the United States. (n.d.). An HSUS Report: the welfare of cows in the dairy industry. Retrieved from http://www.humanesociety.org/assets/pdfs/farm/hsus-the-welfare-of-cows-in-the-dairy-industry.pdf.

[122] Lean IJ, Annison F, Bramley E, Browning G, Cusack P, Farquharson B, et al. Ruminal acidosis prevention and treatment. (June), 60. Retrieved from, https://www.ava.com.au/sites/default/files/documents/Other/RAGFAR_doc.pdf; 2007.

[123] Humane Society International. Fast facts on veal crates in Canada: Humane Society International. Retrieved from http://www.hsi.org/world/canada/work/intensive-confinement/facts/veal-crates-canada-facts.html; 2018.

[124] McKenna C. The case against the veal crate. Retrieved from, https://www.ciwf.org.uk/media/3818635/case-against-the-veal-crate.pdf; 2001.

[125] Solano J, Orihuela A, Galina CS, Aguirre V. A note on behavioral responses to brief cow-calf separation and reunion in cattle (Bos indicus). J Vet Behav Clin Appl Res 2007;2(1):10–4. https://doi.org/10.1016/j.jveb.2006.12.002.

[126] Wagner K, Seitner D, Barth K, Palme R, Futschik A, Waiblinger S. Effects of mother versus artificial rearing during the first 12 weeks of life on challenge responses of dairy cows. Appl Anim Behav Sci 2015;164:1–11. https://doi.org/10.1016/j.applanim.2014.12.010.

[127] Loret de Mola C, Horta BL, Gonçalves H, Quevedo LA, Pinheiro R, Gigante DP, et al. Breastfeeding and mental health in adulthood: a birth cohort study in Brazil. J Affect Disord 2016;202:115–9. https://doi.org/10.1016/j.jad.2016.05.055.

[128] Thornton PK. Livestock production: recent trends, future prospects. Philos Trans R Soc Lond B 2010;365(1554).

[129] Gura S. Industrial livestock production and its impact on smallholders in developing countries. Retrieved from, https://foodsecurecanada.org/sites/foodsecurecanada.org/files/Industrial_livestock_prod_devel_countries.pdf; 2009.

[130] Demeter Inc. Demeter Association Biodynamic Processing Standards. Retrieved from, http://www.demeter-usa.org/downloads/Demeter-Processing-Standards.pdf; 2017.

[131] Government of Canada. Canadian General Standards Board Organic production systems General principles and management standards. Retrieved from, https://www.cog.ca/wp-content/uploads/2018/05/032_0310_2015-e_Amended-in-2018.pdf; 2018.

Glossary

Apolipoprotein A1 a protein which is part of high density lipoprotein cholesterol and facilitates removal of cholesterol from the arterial wall back to the liver. The the ratio of Apoliporotein B to A1 is used an an indicator of risk for coronary heart disease.

Apolipoprotein B a protein which is part of low density lipoprotein cholesterol (as well as VLDL, IDL, and chylomicrons). The role is not entirely certain although it known as an essential organizing protein within these structures. The ratio of Apolipoprotein B to A1 is used an indicator of risk for coronary heart disease.

All-cause mortality the total number of deaths due to all causes of death over a specific period of time.

Beta glucan a type of soluble fiber that has been shown to reduce serum cholesterol levels. It is found in foods such as oats and barley.

Bile a brown/yellow fluid produced by the liver and stored in the gall bladder. After a meal, the gall bladder discharges the bile into the duodenum to help with digestion of fats and absorption of fat soluble vitamins. It is made up of bile salts, cholesterol, phospholipids and bilirubin. Most of the bile salts are re-absorbed in the ileum, while some are lost through the feces.

Blood Pressure Systolic Blood pressure is the maximum pressure during one heart beat while the diastolic pressure is the minimum pressure during one heartbeat. Normal resting blood pressure is 120/80mmHg (systolic/diastolic).

Body Mass Index (BMI) is an individuals' weight in kilograms divided by their height in meters squared. This calculation helps standardize the natural differences you see in body weight for a short person compared to a tall person.

Carbohydrate (carb) two main types of absorbable carbohydrates are present in foods: sugars (e.g. monosaccharides: glucose, fructose, galactose, and disaccharides: maltose, sucrose and lactose) and starches (amylose and amylopectin). The starches are broken down into glucose by the body and absorbed into the bloodstream. Non-absorbable carbohydrates include dietary fiber, which is not subject to digestion by endogenous enzymes but may be digested by bacteria in the large intestine.

Carcinogen a substance with the potential to cause cancer.

Cardiovascular disease (CVD) includes diseases that involve the heart and blood vessels, with outcomes such as myocardial infarction, stroke, peripheral vascular disease, angina, etc.

Cholesterol (dietary) refers to the cholesterol found in food and is different than endogenous cholesterol which is produced by the body. Dietary cholesterol is found in animal products such as meat, fowl, eggs, and dairy. Dietary cholesterol is not found in plant products.

Cholesterol (ester) an ester of cholesterol associated with atherosclerosis.

Cholesterol (serum) is the lipid molecule present in the serum. Cholesterol is an essential component of cell membranes. In humans, sufficient cholesterol is synthesized by the liver (endogenous cholesterol) and no dietary source is required. Cholesterol also serves as a precursor to several hormones, bile acids and vitamin D.

Chylomicrons are large triglyceride-rich particles that carry triglycerides, cholesterol and fat soluble vitamins from the intestines, via intestinal lymphatics, to the circulation. They are associated with a variety of apolipoproteins, including A-I, A-II, A-IV, B-48, C-I, C-II, C-III, and E.

Coronary heart disease (CHD) includes only diseases of the coronary arteries such as myocardial infarction, angina, etc.

Cohort studies a study that follows participants over time to assess various health outcomes.

Complete protein contains all nine essential amino acids required in the human diet. Soybeans and quinoa are examples of plant-based sources of complete proteins.

Concentrated feedlots a system of farming where animals are fed high-energy diets in order to gain mass quickly. Feed is typically grown elsewhere. These are also known as factory farms or concentrated feedlot operations.

Control Group a group of individuals in a study that are identical to the experimental group but have not been subject to any treatment or intervention. Their outcomes are compared against those of the treatment group to determine whether any effect of treatment exists.

C-reactive protein (CRP) is an acute phase protein measured in plasma as a marker of inflammation. Increased levels of CRP have been associated with increased risk of diabetes, heart disease and hypertension.

Dietary fiber carbohydrates and lignin that are of plant origin and are not digested and absorbed by the small intestine

Dietary Guidelines for Americans (DGA) a set of guidelines to help Americans eat healthier diets. Mainly intended for use by policy makers and health professionals. The most current edition (2015-2020) offers 5 overarching guidelines.

Dietary Guidelines Advisory Committee 2015 (DGAC) group composed of nationally recognized medical researchers, academics and practitioners, responsible for reviewing the body of literature for development of new dietary guidelines.

Dietary Patterns are defined by the DGAC in 2015 as "the quantities, proportions, variety or combinations of different foods and beverages in diets, and the frequency with which they are habitually consumed".

Dietary Portfolio/Portfolio Diet plant-based diet designed to lower LDL-C and other CVD risk factors. The four main dietary components include: plant-based protein, plant sterols, viscous fiber and nuts/seeds.

Enterocyte cells intestinal absorptive cell.

Evidence based diet a diet proven scientifically to be effective in achieving an end point.

Food and Agriculture Organization (FAO) a specialized branch of the United Nations with the purpose of eliminating world hunger.

Framingham Risk Equation predicts the 10 year cardiovascular risk of an individual. Different calculators are available depending on the health status of the individual. Most include, sex, age, blood pressure, total to HDL-Cholesterol ratio, smoking, and presence of diabetes.

Glycemic index (GI) is the relative ranking of carbohydrate in foods according to how much they raise the glucose level in the blood. It is a measure of the quality of the carbohydrate.

Glycemic load (GL) is based on both the quality (GI) and quantity of the carbohydrate in a food.

Greenhouse Gas (GHG) is a gas that absorbs infrared radiation, contributing to the greenhouse effect or warming effect. Examples of greenhouse gases include carbon dioxide, methane and nitrous oxide.

High density lipoprotein (HDL-C) particles that carry cholesterol esters and are considered cardio protective because they are thought to be responsible for reverse cholesterol transport, i.e. taking cholesterol from the arterial wall back to the liver. The apolipoproteins (apo) A-I, A-II, C-I, C-II, C-III, D, and E are associated with HDL particles.

Hypercholesterolemia a type of hyperlipidemia, characterized by an excess of cholesterol in the bloodstream.

Hyperlipidemia characterized by excess fats or lipids in the bloodstream.

Industrial agriculture is a modern method of agriculture where animals such as pigs, chickens, cows and fish are raised partially or entirely on feed grown elsewhere. These are also known as factory farms or concentrated feedlot operations.

Large intestine or colon is the last part of the digestive tract and is joined to the terminal ileum via the ileocecal valve. It is responsible for the absorption of water, short-chain fatty acids, minerals vitamin K and a wide range of other compounds. It also contains the microbiome (bacteria, etc.) the makeup of which is now thought to influence many physiological pathways.

LDL-C/HDL-C ratio the ratio of LDL-C and HDL-C particles in the blood. An increased ratio is associated with increased CVD risk.

Lovastatin a drug used to treat high cholesterol levels.

Low density lipoprotein (LDL-C) particles that carry cholesterol esters. It is considered to be atherogenic. The apolipoproteins B-100 and C-III are associated with LDL-C.

Meat analog a plant-based product made to mimic meat in texture and taste.

Mediterranean diet is a diet based on typical foods eaten in the Mediterranean region and consists mostly of fruits, vegetables, legumes, nuts, olive oil and whole grains, fish and small amounts of lean meats.

Meta-analysis a study where data from several independent studies on the same topic are combined to determine if there is an overall trend or effect of treatment.

Metabolic trials trials where all the food is provided to the subjects over the duration of the study.

Metabolic syndrome conditions that co-occur, increasing the risk of CVD and type 2 diabetes. For example: high blood pressure, high cholesterol or triglyceride levels, high blood glucose, abdominal obesity, etc.

Micelle a lipid that arranges itself in a spherical form in an aqueous solution.

Monounsaturated Fatty Acids (MUFA) are fatty acids that have one double bond, food sources of MUFA's include olive oil, olives, avocado, and many nuts.

Myocardial infarction another term to describe a heart attack. It occurs when the blood supply to a part of the heart slows or stops causing damage to the heart.

Oxidized LDL-C is LDL-C damaged by free radicals producing a harmful type of cholesterol.

Phytochemicals are found in plants and used to aid plants in fighting pathogens, predators and increasing their competitive ability. Many health benefits have been attributed to their consumption.

Plant-Based diet which includes only plant-based foods and does not contain any animal-based foods such as dairy, meat or fish.

Plant Sterol phytosterols (sterols and stanols) are naturally occurring substances in the plant wall and are similar to cholesterol in human cells. They occur naturally in small quantities in plant foods such as leafy vegetables, nuts, seeds, and vegetables oils. Normal intakes range from about 200-300 mg/day while vegetarian diets can contain upward of 700 mg/day.

Polyunsaturated fat (PUFA) are fatty acids that contains more than one double bond, of which linoleic and linolenic acid are considered essential to the human body. Examples of foods containing PUFA's are vegetable oils, walnuts, flax and chia.

Pulses are the edible seeds of legumes such as chickpeas, lentils, kidney beans and navy beans.

Saturated Fatty Acids (SFA) fatty acids without any double bonds. Food sources of SFA's include beef, lamb, lard, butter and egg yolks.

Short Chain Fatty Acids (SCFA) are derived from the fermentation of indigestible foods in the large intestine, e.g. acetone, propionate and butyrate.

Small intestine joins the stomach and the colon (large intestine) and consists of the duodenum, jejunum and ileum. The average length is about 6m (20ft). The duodenum is lined by absorbative cells covering villus (finger-like structures) that line the small intestine. It receives digestives juices from the pancreas and bile from the liver to help digest food.

Textured Vegetable Protein (TVP) is typically made from defatted soy flour formed into soy meat or chunks and is often used as a meat substitute.

Trimethylamine- N-oxide (TMAO) occurs when substances (choline and carnitine) found in animal products, especially meats, undergo bacterial fermentation in the gut.

Vegan Diet includes only plant-based foods and does not contain any animal based foods such as dairy, meat or fish.

Vegetarian Diet plant-based diet which may also include dairy products (lacto vegetarian) and/or eggs (ovo vegetarian).

Very low density lipoprotein (VLDL) particles which carry endogenously synthesized triglycerides and to a lesser degree cholesterol and are considered atherogenic. The major apolipoproteins associated with VLDL are B-100, C-I, C-II, C-III, and E.

Viscous Fibers are dietary fibers, which when mixed with water, form a viscous solution, examples of viscous fibers are pectin, guar gum, psyllium, and konjac mannan. Foods rich in viscous fibers include oats, barley, psyllium containing breakfast cereals, eggplant and okra.

Whole grain contains the endosperm, the bran and the germ of the grain, it includes grains such as oats, barley, rye, corn, quinoa, etc.

Whole wheat contains the endosperm, the bran and the germ of the wheat kernel.

Index

Note: Page numbers followed by *f* indicate figures, *t* indicate tables, and *b* indicate boxes.

Printed in the United States
By Bookmasters